BIOMECHANICAL EVALUATION OF MOVEMENT IN SPORT AND EXERCISE

Now in a fully updated and revised new edition, this is still the only up-to-date, practical guide to the use of technology in sport and exercise biomechanics. It includes detailed explanations of the key theory underlying biomechanics testing and measurement, along with advice on choosing equipment and using it effectively.

The second edition includes two completely new chapters on qualitative movement analysis and the assessment of movement coordination, and covers every key functional area in the biomechanics curriculum, including:

- motion analysis using video and on-line systems
- measurement of force and pressure
- measurement of torque and power using isokinetic dynamometry
- electromyography
- computational simulation and modelling of human movement
- research methodologies
- data processing.

Published in association with the British Association of Sport and Exercise Sciences (BASES), it includes contributions from world leading researchers and pioneers in the field of sport and exercise biomechanics. *Biomechanical Evaluation of Movement in Sport and Exercise* is a must-have text for all biomechanics laboratories and for any student undertaking a research project or course in methods, measurement or analysis in biomechanics.

Carl J. Payton obtained his PhD in biomechanics from Manchester Metropolitan University, UK, in 1999 where he is now a Reader and leads the biomechanics and long-term conditions research group. Carl's main research interests are in the biomechanical determinants of elite swimming performance and in the classification of physically impaired swimmers for competition. He has provided biomechanics support to the GB Para-Swimming team since 2000. Carl has previously been Biomechanics Section Chair and member of the BASES Executive Committee.

Adrian Burden began his career at Brunel University before moving to Brighton University and then to Manchester Metropolitan University, UK, in 2002. He obtained his PhD in surface electromyography in the same year and is Associate Head of the Department of Sport and Exercise Sciences. Adrian is also a Reader in Biomechanics and a Senior Fellow of the Higher Education Academy. His main interests lie in the application of electromyography to exercise, clinical and sport settings, and he has run workshops for BASES in this area.

BIOMECHANICAL EVALUATION OF MOVEMENT IN SPORT AND EXERCISE

The British Association of Sport and Exercise Sciences Guide

2nd Edition

Edited by Carl J. Payton and Adrian Burden

Routledge
Taylor & Francis Group
LONDON AND NEW YORK

BASES
The British Association of Sport and Exercise Sciences

Reprint 2019

Second edition published 2018
by Routledge
2 Park Square, Milton Park, Abingdon, Oxon OX14 4RN

and by Routledge
711 Third Avenue, New York, NY 10017

Routledge is an imprint of the Taylor & Francis Group, an informa business

First edition published by Routledge 2008

British Library Cataloguing-in-Publication Data
A catalogue record for this book is available from the British Library

Library of Congress Cataloging-in-Publication Data
Names: Payton, Carl, editor. | Burden, Adrian, editor. | British Association of Sport and Exercise Sciences.
Title: Biomechanical evaluation of movement in sport and exercise : the British Association of Sport and Exercise Sciences guide / edited by Carl Payton and Adrian Burden.
Description: 2nd edition. | Abingdon, Oxon ; New York, NY : Routledge, 2018. | Includes bibliographical references and index.
Identifiers: LCCN 2017034261
Subjects: LCSH: Human mechanics. | Exercise—Physiological aspects. | Sports—Physiological aspects.
Classification: LCC QP303 .B5585 2018 | DDC 612.7/6—dc23
LC record available at https://lccn.loc.gov/2017034261

ISBN: 978-0-367-24193-3 (pbk)
ISBN: 978-0-203-09554-6 (ebk)

Typeset in Sabon
by Apex CoVantage, LLC

Printed and bound in India

For sale in India, Pakistan, Nepal, Bhutan, Bangladesh and Sri Lanka only.

CONTENTS

ILLUSTRATIONS

TABLES

FIGURES

CONTRIBUTORS

Vasilios Baltzopoulos is the Head of the Research Institute for Sport and Exercise Sciences (RISES) at Liverpool John Moores University. He is the lead author of the BASES expert position statement on assessment of muscle strength with isokinetic dynamometry and the organiser of the BASES workshops in these areas. He has served as Biomechanics Section Chair and member of the BASES Executive Committee and Editor of the Biomechanics Section of the *Journal of Sports Sciences*.

Roger M. Bartlett retired from his position as Professor of Sports Biomechanics in the School of Physical Education, University of Otago, New Zealand, in 2014. He is an Invited Fellow of the International Society of Biomechanics in Sports (ISBS) and European College of Sports Sciences, and an Honorary Fellow of the British Association of Sport and Exercise Sciences, of which he was Chairman from 1991–1994.

Adrian Burden began his career at Brunel University before moving to Brighton University and then to Manchester Metropolitan University in 2002. He obtained his PhD in surface electromyography in the same year and is Associate Head of the Department of Sport and Exercise Sciences. Adrian is also a Reader in Biomechanics and a Senior Fellow of the Higher Education Academy. His main interests lie in the application of electromyography to exercise, clinical and sport settings, and he has run workshops for BASES in this area.

John H. Challis is a Professor at the Pennsylvania State University. He obtained his BSc (Honors) and PhD from Loughborough University of Technology. He is a fellow of the National Academy of Kinesiology (US) and the American Society of Biomechanics, and has been a President of the American Society of Biomechanics and the International Society of Biomechanics. His research focuses on the coordination and function of the musculoskeletal system, and data collection and processing methods.

Nachiappan Chockalingam is a Chartered Engineer, a Chartered Scientist and a Principal Fellow of the Higher Education Academy. His doctoral thesis in 2004 at Staffordshire was on gait and posture analysis of patients with scoliosis. As Professor of Clinical Biomechanics, he leads biomechanics research at Staffordshire. He is also an Affiliate Professor at the Faculty of Health Sciences, University of Malta, and a Visiting Professor at the Department of Sports Medicine and Arthroscopy, Sri Ramachandra University, Chennai.

Aoife Healy is a Senior Research Officer within the Clinical Biomechanics Team at Staffordshire University. Her research interests include gait analysis and plantar pressure measurement. She joined Staffordshire University as a KTP Research Associate in 2009 and completed her PhD in Biomechanics at that university. She graduated from the University of Limerick, Ireland, with a BSc in Sports and Exercise Sciences and completed her Masters in Biomechanics at Dublin City University, Ireland.

Christopher R. Hudson completed his Sports Engineering PhD at Sheffield Hallam University, UK, in 2015. Chris has developed a wide range of systems to help monitor elite athletes' training and tournament performances. He has over 20 years of software programming experience and has an interest in building cloud- and mobile-based applications that collect, store and analyse data about an athlete's performance. Chris's main research area is in the application of novel photogrammetry techniques.

Mark A. King is a Reader in Sports Biomechanics at Loughborough University, UK, specialising in using subject-specific computer simulation models to understand optimum performance in sport. He has been at Loughborough since 1990, graduating in Sport Science and Mathematics in 1993 and obtaining his PhD in computer simulation of dynamic jumps in 1998. In addition he is the current chair of International Society of Biomechanics Technical Group on Computer Simulation.

Peter F. Lamb is a Lecturer of Sports Biomechanics in the School of Physical Education, Sport and Exercise Sciences at University of Otago, New Zealand. He is also co-coordinator of the Sports Technology major in the BAppSc programme. His research interests are in sports performance analysis as well as analysis methods for assessing coordination.

Adrian Lees recently retired as Professor of Biomechanics at Liverpool John Moores University. He received his PhD in Biomechanics from the University of Leeds in 1977. He is currently Emeritus. In 2003 he was awarded Doktor Honoris Causa from the Academy of Physical Education, Warsaw. In sport biomechanics, he has a particular interest in sport technique and its application to soccer. He has also worked as a consultant to the British Athletics Federation, focussing on the Long and Triple Jump events. In rehabilitation biomechanics, he has run research programmes in wheelchair performance and amputee gait. Adrian has authored over 15 books and book chapters, and over 120 peer-reviewed scientific research papers.

Clare E. Milner is an Associate Professor in the Department of Physical Therapy and Rehabilitation Sciences at Drexel University, Philadelphia, PA, USA, where she specializes in applied biomechanics. Her research interests focus on the biomechanics of lower extremity injury and rehabilitation, in particular the occurrence of overuse injuries in runners and gait in people with knee pathology.

David R. Mullineaux was appointed as Professor in Sports Science in 2011 to shape the research agenda of the University of Lincoln's emerging School of Sport and Exercise Science. He has made several transitions between academia and industry in the UK and USA gaining experience of teaching, consulting and researching in biomechanics and research methods. His research interests are in using real-time biofeedback to alter technique, and on applying analytical techniques to sports biomechanics data.

Carl J. Payton obtained his PhD in Biomechanics from Manchester Metropolitan University, UK, in 1999 where he is now a Reader and leads the biomechanics and long-term conditions research group. Carl's main research interests are in the biomechanical determinants of elite swimming performance and in the classification of physically impaired swimmers for competition. He has provided biomechanics support to the GB Para-Swimming team since 2000. Carl has previously been Biomechanics Section Chair and member of the BASES Executive Committee.

Jonathan Wheat is a Reader in Biomechanics in the Centre for Sports Engineering Research at Sheffield Hallam University, UK. Jonathan leads their Biomechanics Research Group and is the coordinator for the Research Excellence Framework. His research interests centre on the application Ecological Dynamics to sport and exercise biomechanics data collection and analysis. This includes work to enable collection of biomechanics data in more representative environments. His work also involves the application of Ecological Dynamics analytical techniques and feeding back biomechanics data in an appropriate manner.

Maurice R. Yeadon graduated in Mathematics from the University of Cambridge in 1968 and after a number of years teaching mathematics obtained his PhD in the computer simulation of aerial movement at Loughborough University in 1985. He then took up a biomechanics position at the University of Calgary and in 1990 returned to Loughborough University where he is currently Emeritus Professor of Computer Simulation in Sport.

CHAPTER 1

INTRODUCTION

Carl J. Payton and Adrian Burden

The British Association of Sport and Exercise Sciences (BASES) is the professional body for sport and exercise sciences in the UK and its mission is to promote excellence in this field (www.bases.org.uk). One of BASES' key roles is to develop and enhance the professional and ethical standards of its members. It achieves this through hosting annual conferences, offering regular CPD workshops, providing professional accreditation schemes and providing practical guidelines on best practice in each of the sport and exercise sciences.

This second edition of *Biomechanical Evaluation of Movement in Sport and Exercise* is in fact the fifth version of guidelines written for sport and exercise biomechanists and endorsed by BASES. These guidelines are still informally known as the 'BASES Biomechanics Guidelines' within the British sport and exercise biomechanics community. This edition marks more than 25 years since publication of the first set of guidelines. Previous versions were edited either solely (Bartlett, 1989; 1992; 1997) or jointly (Payton and Bartlett, 2008) by Roger Bartlett, who decided to step down as editor for this edition. We would like to thank Roger for instigating the guidelines and for his hard work in editing and contributing to them over the years.

Texts that focus on basic biomechanical principles in sport and exercise tend to remain up-to-date and relevant for considerably longer than texts, such as this, which focus on the use of biomechanics technology. Since the publication of the first edition of this book in 2008, there have been some significant advances in the technologies used in biomechanical measurement. This revised text is necessary to reflect these changes. Additionally, some of the content in the previous edition has become almost obsolete. For example, cameras that record to videotape are likely to become redundant within the next few years and therefore warrant only a mention in the second edition.

OVERVIEW OF CONTENT

This book contains new versions of each chapter from the previous edition, covering topics that remain as relevant today as they were nine years ago. Two new chapters have also been included in this edition. The first is *Qualitative Biomechanical Analysis of Technique*, written by Adrian Lees. This chapter provides the basis of qualitative analysis, starting with analysis models, continuing with principles of movement, and concluding with some contemporary thoughts and developments on this topic. The second is *Assessing Movement Coordination*, written by Peter Lamb and Roger Bartlett. Here, the various qualitative and semi-quantitative methods available for describing and analysing coordination patterns are covered. The chapter also considers how coordination patterns can add substantially to our understanding of sport and exercise movements, and help identify where changes might be made to improve performance. Both of these chapters are invaluable additions to the book given that an increasing number of sport and exercise biomechanists are interested in the study of movement coordination, and the majority of sports biomechanists working with performers often analyse technique qualitatively as well as quantitatively.

Chapters from the previous version of this book have been updated and, in some cases, substantially re-written to reflect the academic and technological advances that have occurred over this period. *Motion Analysis Using Video* by Carl Payton and Christopher Hudson has been re-written to reflect recent developments in video hardware and software. New material on methods of remotely capturing video data, display hardware and video formats have been included. Alternative calibration methods for two- and three-dimensional video analysis are also now presented. *Motion Analysis Using On-line Systems*, by Clare Milner, has been updated and re-ordered to reflect improvements in technology and developments in volume calibration. Nachiappan Chockalingam and Aoife Healy have produced a substantially updated chapter titled *Measurement of External Forces*. In *Surface Electromyography*, Adrian Burden provides new advice on how choices made when processing EMG data affect reliability and statistical significance of derived variables. He also provides additional guidance on how to normalise EMGs, following publication of a number of review articles in this area since the last edition. The effect of misalignment between the axis of the joint and the dynamometer on the accuracy of moments measured by dynamometers is re-evaluated by Vasilios Baltzopoulos in *Isokinetic Dynamometry*. In *Data Processing and Error Estimation*, John Challis has expanded his section on methods of calculating joint angles three-dimensionally and the errors associated with the associated computational procedures. David Mullineaux and Jonathan Wheat's chapter on research methods now includes an appraisal of statistical techniques used to analyse multiple biomechanical variables and continuous sets of data, in addition to discrete points in a time series. Fred Yeadon and Mark King have upgraded the final chapter, *Computer Simulation Modelling in Sport*, with additional material and well over 30 new references.

This book, as the title suggests, is intended to provide practical guidance, with theoretical underpinning, to those involved in the measurement, processing and analysis of biomechanical information, particularly in the sport and exercise domains. It should therefore be of particular relevance to applied sport and exercise biomechanists, clinical biomechanists, researchers, performance analysts and educators. The book should also serve as a valuable resource for students studying biomechanics at both undergraduate and postgraduate level.

Carl Payton and Adrian Burden, June 2017

REFERENCES

Bartlett, R. M. (1989) *Biomechanical Assessment of the Elite Athlete*, Leeds: British Association of Sports Sciences.

Bartlett, R. M. (ed.) (1992) *Biomechanical Analysis of Performance in Sport*, Leeds: British Association of Sport and Exercise Sciences.

Bartlett, R. M. (ed.) (1997) *Biomechanical Analysis of Movement in Sport and Exercise*, Leeds: British Association of Sport and Exercise Sciences.

Payton, C. J. and Bartlett, R. M. (eds) (2008) *Biomechanical Evaluation of Movement in Sport and Exercise – The British Association of Sport and Exercise Sciences Guidelines*, Oxon: Routledge.

CHAPTER 2

QUALITATIVE BIOMECHANICAL ANALYSIS OF TECHNIQUE

Adrian Lees

INTRODUCTION

Qualitative analysis is a method used to evaluate technique in the performance of sports (or exercise) skills. It uses observation and can be supplemented with a visual recording, such as video, with playback facilities such as single frame, slow motion and repeated viewing. It relies on a knowledge of the relevant sport and sports skill, as well as a knowledge of 'principles of movement'. Finally, qualitative analysis should be conducted using a chosen analysis model.

It is important to distinguish between technique and performance. Technique is defined in general terms as the 'way of doing' and the technique used in the performance of a specific sports skill can be defined as the way in which that sports skill is performed. Performance is defined in terms of outcome, and technique is only one of several factors that influence outcome. This distinction has been difficult for biomechanists to make due to the popularity in recent years of quantitative analysis and attempts by investigators to undertake a 'biomechanical analysis of performance' which, in many cases, attempts to measure aspects of technique. A simple example can illustrate the difference. In a sprint, an individual may be able to demonstrate the technique (way of doing) of an elite sprinter but not be able to achieve the same level of performance. This is because the individual lacks the physical, physiological, psychological or other attributes required for a higher-level performance. Thus, good performance requires good technique, but good technique does not guarantee good performance.

Good technique is a prerequisite of good performance so it is sensible to attempt to understand what good technique means for individual sports skills. The advantage of the qualitative analysis approach is that an analysis can be undertaken quickly, without recourse to expensive, time consuming and often restrictive methods characterised by contemporary quantitative methods. A further advantage is that a sports skill can be analysed holistically; something that is quite difficult to do using quantitative methods.

Qualitative analysis is not a new approach to the analysis of sports skills. In fact this was the main approach in the 1950s and 1960s, mainly due to increasing interest in the techniques used to perform sports skills but the lack of equipment and methods to investigate them (see Lees, 2002, for a review). Out of these early studies evolved what we now recognise as qualitative analysis. Unfortunately, these approaches were overtaken by the thrust in biomechanics for quantitative analysis; a suitable framework for qualitative analysis was never developed. Researchers have subsequently defined their own frameworks but no generally agreed approach has emerged. In this chapter I provide the basis of qualitative analysis, starting with analysis models, continuing with principles of movement, and concluding with some contemporary thoughts and developments on this topic.

ANALYSIS MODELS

The phase analysis model

Phase analysis is the descriptive process of dividing up a movement into relevant parts so that attention can be focused on the technique used in each part. Some authors identify three main phases to a skill (*retraction, action and follow-through*), while others identify four main phases (those above but preceded by a *preparation phase*), a recognition that the start of a movement may also influence the way a movement is performed (e.g. in a soccer kick the distance away from the ball and angle of approach are relevant factors to performance; in golf, the stance, how the club is held and position of the feet relative to the ball are also relevant factors). Some authors identify more than four phases but these are often sub-phases of the four main phases mentioned earlier. Most authors acknowledge that the phases can be further broken down into *sub-phases* and that the distinction between one phase or sub-phase and another is arbitrary and determined by the particular skill and the needs of the analyst. Nevertheless, this process of breaking a skill down into its functional parts is an important first analytical step.

The *preparation phase* describes the way in which the performer sets or prepares for the performance of the skill. For example, as noted earlier, it may relate to the start position and/or the way the ball is placed in a soccer penalty kick, or the way the club is held and/or ball placed on the tee in a golf drive. The *retraction phase* refers to the withdrawal of, typically, the arm or leg prior to beginning the main effort of performance. For example in kicking, the kicking leg is drawn back; in tennis the retraction phase is represented by the backswing. The *action phase* is where the main effort of the movement takes place. For example, forward motion in executing a tennis serve, or the throwing action when propelling a javelin. The *follow-through phase* allows the movement to be slowed down under control and is thought to be necessary to avoid injury that might occur with rapid limb deceleration. Some examples are given in Table 2.1. As noted earlier, the phase analysis model requires the phases and sub-phases, if appropriate, to be identified and described. Some examples of sub-phases are given in Table 2.2.

Table 2.1 Examples of the four phases from selected skills.

Event/skill	*Preparation*	*Retraction*	*Action*	*Follow-through*
Long jump	Run up	Body adjustments during penultimate stride to touchdown	Take-off, from touchdown to take-off	Flight and landing
Golf swing	Correct grip and stance	Backswing	Downswing	Follow-through
Penalty kick	Run up	Retraction of kicking leg	Forward swing of kicking leg and contact	Follow-through

Table 2.2 Examples of the sub-phases for the action phase of selected skills.

Event/skill	*Action phase*	*Sub-phase 1*	*Sub-phase 2*	*Sub-phase 3*
Long jump	Take-off	Compression (knee flexion)	Extension (knee extension)	-
Golf swing	Downswing	Weight shift initiation of movement with the hips	Wrist locked arms rotate as single unit	Wrist unlocks and arms rotate as double unit
Penalty kick	Forward swing of kicking leg	Hips rotate forwards and leg flexes at knee	Knee extends to contact	-

The description of phases and sub-phases should identify *key moments* and *critical features*. Key moments are those points in time at which an important action is performed related to the 'way of doing'. One important key moment in striking sports is impact. However, events such as toe-off, foot-strike, maximum knee flexion, or minimum elbow angle, to name a few, would all be actions that define a key moment of the technique. Critical features are observable aspects of a movement, and refer to body or limb position (e.g. when catching a ball – crouch with arms and legs flexed; for a tennis serve – the 'back-scratch' position of the racket in the backswing) and motions (e.g. when catching a ball – give or retract the hands with the ball). It is worth noting that critical features are often related to coaching points and often are expressions of selected underlying movement principles (dealt with later). To complete the phase analysis model, the movement principles associated with the phase or sub-phase, key moments and/or critical features need to be identified. Movement principles are dealt with separately later in the chapter. The phase analysis model is given schematically in Table 2.3.

The performance outcome model

An alternative approach to the phase analysis model involves an analysis of the factors that influence performance. By focusing on those factors, it is claimed

Table 2.3 Schematic template for a phase analysis model.

Phase description	*Preparation*	*Retraction*	*Action*	*Follow-through*
Sub-phase description[1]				
Key moments[2]				
Critical features[3]				
Movement principles[4]				

1 Each box under each phase should contain a brief description of the phase or (if appropriate) the sub-phase.
2 Key moments are often related to the start and end of a phase or sub-phase.
3 Critical features are often related to key coaching points.
4 Movement principles can be identified by their abbreviated title (see text).

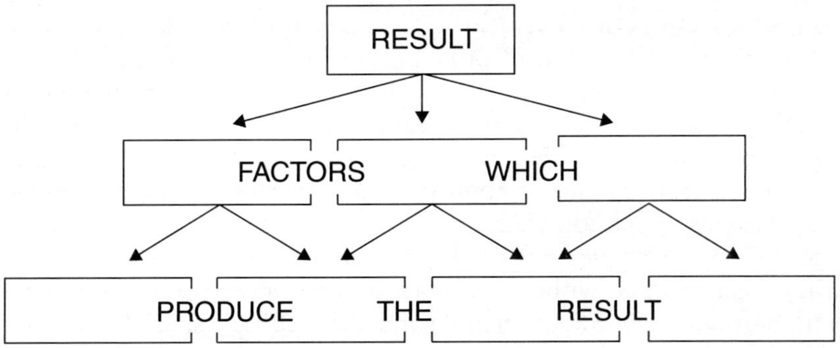

Figure 2.1 The Hay and Reid performance outcome model.

that faults and limiting factors in performance can be identified. The most influential of these models was proposed by Hay and Reid (Hay and Reid, 1982). The model was first developed for use in qualitative analysis but has also found widespread use in quantitative analysis to assist biomechanists in identifying important variables for quantification (Chow and Knudson, 2011). The model does not address aspects of technique (the way of doing) directly, but the mechanical relationships that govern performance. In that sense, the model is more closely linked with movement principles and one may view the model as a more direct and systematic approach to the identification of the mechanical principles that govern performance. The model is constructed as a hierarchy of factors on which the result (outcome) of the performance is dependent (Figure 2.1). The rule that governs the construction of a model for a particular skill is that each of the factors in the model should be completely determined by those factors that appear immediately below it either by (1) addition or (2) known mechanical relationships.

As noted earlier, the performance outcome model does not address issues of technique directly. For example, in a golf drive, the model will tell us that the speed of the club head must be high at impact, but not how to achieve it. Information on how to use the arms and club as a two-lever system, weight shift and hip-shoulder rotation are beyond the scope of the model. In other words, the model is able to identify factors relevant to performance but not aspects of technique relevant to these factors. Nevertheless, the Hay and Reid model would appear to be valuable for identifying a range of factors influencing performance and providing a framework from which technique can be discussed. In this sense, it is not an alternative model for technique analysis but is complementary to other methods.

MOVEMENT PRINCIPLES

The earliest and perhaps the most widely used scientific approach to the evaluation of technique is the application of *mechanical* principles. These have been articulated and developed over time and have tended to be referred to as 'biomechanical principles of movement' or *movement principles* for short. In general, these are a combination of principles based on mechanical relationships, multi segment interactions and biological characteristics of the human musculoskeletal system. A movement principle is a description of how to achieve a specific movement outcome based on sound mechanical and/or biological principles. A movement principle is applied, in general terms, to help to understand how sports skills are performed.

Movement principles can be classified according to the general outcomes that they are associated with. These outcomes are speed production, force production, movement coordination and some that relate to special circumstances. Speed and force principles are based on mechanical relationships, such as those described by Newton's second law, and principles, such as the conservation of momentum. The equations underpinning these principles are not detailed here, to preserve the 'qualitative' nature of this chapter, but can be found in any good biomechanics text. The coordination principles are based on multi-segment interactions and have mechanical or biological foundations. This group of principles reflect the more complex operation of the human body. The inclusion of biological principles, such as the stretch-shorten cycle, is an acknowledgement of the biological factors that determine complex human movement. This is an area where a further development of movement principles would be warranted. The principles applied to special circumstances are where a number of interacting mechanical factors are commonly encountered, such as in the flight of projectiles, or where other phenomenon are frequently observed, for example the speed-accuracy trade off, reflecting a movement control limitation. There is no general agreement as to the number, or even the names, of movement principles. It is possible to identify principles that are related to specific performances. Some authors have identified as many as 53 movement principles, while others as few as six (Lees, 2002). The higher number tends to reflect the specific mechanical principles, whereas the lower number tends to reflect the more general principles (e.g. Bartlett, 2007). There have been some attempts to reduce the larger number of principles to a manageable form as 'core concepts' (Knudson, 2002); these contain a mixture of mechanical, multi-segment and biological principles

Table 2.4 Table of movement principles with codes.

Speed	*Force*	*Coordination*	*Specific Performance*
S1	F1	C1 AR	P1
S2	F2 ROM	C2 PDS	P2
S3	F3	C3	
S4 EPS	F4	C4 SSC	
	F5		
	F6		
	F7		

and provide some use in a practical context. Before illustrating how these are applied, it is necessary to identify what they are and give examples of how each can be used. Following is a list of those principles relevant to most sports and suitable for a qualitative biomechanical analysis of performance; they are outlined in Table 2.4. Although not definitive, they are thought to cover most applications.

Speed (S) principles

Speed principles are a range of movement principles which relate to the generation of speed in a part of or the whole of the body.

(S1) Whole body running speed

Running speed is increased gradually through sequential drives of the legs. The increase in speed, with each stride, is greatest at low speeds and reduces as whole body speed increases. Maximum running speed is achieved after about 40–50 m of sprinting. Therefore, to reach maximum speed, a person must be able to sprint for at least this distance. Some consequences of this are: (1) running speed is often controlled by the performer, due to the complexity of the skill and/or the high forces involved (in these cases the speed is kept sub-maximal in order to complete the skill), and (2) in field games, players are unlikely to reach their top running speed over a 10–15 m sprint, therefore their ability to accelerate is important.

(S2) Whole body rotational speed

Rotational movements of the whole body are completed more rapidly by bringing the limbs closer to the body's axis of rotation. For example, in a somersault the trampolinist rotates more rapidly when tucked; in a pirouette, the ice skater rotates more rapidly when the arms are brought close to the body; in a squash backhand stroke, the action phase begins with the racket arm close to the body so that the whole body can rotate into the movement before extending the racket arm. Conversely, extending the limbs slows the body down. This is done not only to slow down the rotation but to allow the performer more time to make a good landing. For example, stretching the arms above the head after a gymnastic vault slows the forward rotation of the gymnast; opening out after a tucked somersault slows the rotation of a trampolinist.

(S3) Limb rotational speed

To rotate a limb (e.g. arms or leg) rapidly requires the limb to be flexed and held close to the body. For example, in sprinting the leg is flexed tightly during the recovery part of the cycle. In a tennis serve the racket arm is flexed to achieve a back scratch position (a critical feature); in golf the arm and club are in a flexed position at the top of the downswing.

(S4) End point speed (EPS)

A high end point speed requires a large distance from the axis of rotation to the end point. Consequently at impact or release, the limb is at or close to full extension. For example, when striking in sports like tennis, golf, baseball and cricket, the action phase begins with the limb and implement held close to the body and as the phase develops, the end point (hand, foot, racket head, club head) is allowed to move away from the axis of rotation, thus increasing its radius of rotation.

Force (F) principles

Force principles are a range of movement principles which relate to the generation of force used to achieve a specific movement outcome.

(F1) Maximum force production

To produce the maximum effective force, a firm base is required against which to push. For example, in throwing events such as shot and javelin, the action phase occurs when the delivery foot is in firm contact with the ground; in jumping events such as long, triple and high jump, a firm surface is always used to push from. Conversely, if the surface moves, the effective force produced is reduced. For example, soft surfaces, such as turf and sand, are more difficult to run and jump on due to the deformation of the surface. In tennis serving, the server is often seen to come off the ground. Is this an exception? Not really. To gain maximum racket head speed the server extends the legs (generating maximal effective force) and it is only after this that the player throws the racket head towards the ball. This combination of vigorous movements directed vertically causes the body to lift off the ground. Other principles are also used in the performance of this skill in order to achieve high end point speed (see C2 and C4 sections that follow).

(F2) Range of motion (ROM)

The greater the limb's range of motion, the longer a muscle force can be applied for. Consequently, there is a possibility of achieving a greater effect by contracting muscle over a greater range of motion. For example, in running, as speed increases, the stride length increases, which occurs due to the greater extension (i.e. range of motion) of the leg during the drive-off. One implication of this principle is that if the joint has greater flexibility, it is likely that this will allow a greater force producing range of motion, leading to enhanced performance. Consequently there is thought to be a performance aspect to flexibility training.

(F3) Change of running direction

A change in direction of motion when running is produced by applying a force at right angles (perpendicular) to the current direction of motion. For example, in a side step or swerve made by a player in field games, the foot is placed so as to maximize the friction force applied to the surface. This friction force should be directed perpendicular to the current direction of motion and *not* towards the intended direction of motion.

(F4) Impact – stationary ball or object

When hitting a stationary ball or object, the implement making the impact must move in the direction it is intended that the ball or object being hit should go. For example, when taking a penalty kick, a goalkeeper can sometimes guess correctly the direction of the ball by carefully watching the motion of the kicker's foot as it moves to strike the ball.

(F5) Impact – moving ball or object

When hitting a moving ball or object, the striking implement must move in such a direction so as to take into account the motion of the moving object. It will always be the case that the direction of the striking implement at impact will be different from that which the ball or object subsequently goes. This divergence is related to the mass and speed of the two objects respectively (and another expression of principle F3). If, when heading a moving ball or making a cricket, baseball or tennis shot, the implement is swung to drive it in the direction of intended motion, the ball will *not* travel in this direction.

(F6) Stability

Objects are more stable if they have a wide base and low centre of mass. For example, in wrestling, at the start of a competition wrestlers spread their legs and lower their centre of mass to provide a stable base. Stability is a 'statics' concept. Beware of 'dynamic stability' which is a 'dynamics' concept beyond the scope of this chapter and probably not appropriate for a qualitative approach.

(F7) Resistance to motion in fluids

Resistance to motion when moving through air or water is reduced by decreasing the area of the body or object presented to the on-coming air or water (known as the cross-sectional area) and making a more streamlined shape. For example, in cycling, handlebars which are lowered and extended forward enable the cyclist to adopt a smaller cross-sectional area and more streamlined shape. Conversely, resistance is increased by increasing the cross-sectional area and making the shape less streamlined. For example in swimming, the area of the hand, and therefore the resistance produced by the arm pull, can be increased by using a hand paddle.

Coordination (C) principles

Coordination principles are a range of movement principles which relate to the coordination of motion between segments so as to achieve maximal or optimal performance.

(C1) Action–reaction: simultaneous movements of opposing limbs (AR)

The movement of one limb or body part helps the movement of the opposite (or contralateral) limb or body part. For example, in walking, running and sprinting, as one leg comes forward the contralateral arm also comes forward; an effective sprint start is one in which the arms drive vigorously to aid the force generation of the opposite leg. A good coaching point is that 'the arms drive the legs'; in hurdling (crossing over the hurdle) the lift of the lead leg is helped by bringing the opposite arm forward as far as possible; in pike movements in gymnastics, trampoline and diving, the performance of the pike is aided if the upper and lower body are brought together at the same time; the movement of one is helped by the movement of the other; when heading a soccer ball the same pike movement is produced which increases the speed with which the head makes contact with the ball.

(C2) Proximal-to-distal sequence of movements (PDS)

This is used when producing high-speed movements. Many skills require a coordinated sequence of rotational movements to achieve a high end point velocity. This is achieved by rotating the large segments close (proximal) to the body first and terminating in the rotation of the segment farthest (distal) from the body. For example, in most throwing/striking/kicking skills the action often starts with a step forward with the leg contralateral to the throwing/striking/kicking arm or leg, which has the effect of opening the hips. It continues with the hips rotating forward and is then followed by the trunk, shoulders and finally the elbow, hand and implement. The rotation speed of the earlier segment is built upon by the next segment, so as to build-up the speed of the end point sequentially. Rotation of the distal segments cause the end point to move away from the axis of rotation, thus increasing the distance of rotation (see speed principle S4).

(C3) Simultaneous joint movements for force/power production

Simultaneous joint movements are used when producing forceful or powerful actions for a linked body segment chain that includes several of the major joints of the body. To ensure that this link system provides a firm base (see force principle F1) it is important that the muscle groups operate simultaneously. Therefore, forceful/powerful movements require muscles about joints to act synchronously. For example, when jumping for height, the hip and knee extensor muscles act simultaneously to generate high force; in the shot put, the drive from the trunk and legs occur simultaneously; in the bench press, the muscles activating the shoulder and elbow joint act together; in the sprint start, the hip, knee and ankle joint extend together.

Detailed biomechanical analysis has shown that in some cases, even though the muscles act synchronously, the joints do not extend synchronously. Typically in the vertical jump, the ankle joint extends after the hip and knee joints generate their main effort. As the ankle joint is weaker than these other two joints, its role is to maintain a firm base for the other stronger joints as they produce their effort. Once this diminishes, the ankle, which is kept in a position of flexion, extends making its contribution to the movement. The sequence of joint motion is a reflection of the relative strength and function of the joints rather than a violation of the principle.

(C4) Stretch–shorten cycle (SSC)

Many actions involve a pre-stretching of the muscles and tendons, which aids performance by enabling the highest muscle forces to be built up at the beginning of the movement. Consider a throw (or jump) in which the starting position is in the squat position. When the muscles begin to shorten, their force develops gradually. As this force develops, the movement velocity increases, as does the speed of shortening of the muscles, which are now less capable of generating force (as determined by the force-velocity relationship for muscle). The net effect is that the movement is less able to increase speed further. If the muscles are pre-stretched by a countermovement, the muscles are fully active at the beginning of the upward movement. This means that as the upward movement begins, they are generating their maximum effort. The muscle and tendon 'stretch' should be quickly followed by the 'shorten' in order to maximize this effect.

Specific performance (P) principles

Specific performance principles are a range of movement principles which identify the underlying factors relevant to specific aspects of performance.

(P1) Flight and projectile motion

An object which moves through the air under the influence of gravity is called a projectile. The key outcome of projectile motion is often the range, but sometimes the height reached and time of flight are important performance measures. The mechanical factors determining projectile motion are the height, angle and speed of release, with the speed of release being the most important. The effects of air resistance can be important in many situations, particularly those where the projectile is light, the relative velocity of the wind is high, or the object is shaped so as to have aerodynamic properties. The flight path of the projectile is modified accordingly and this as well as more complex effects, e.g. spin, need to be taken into account.

(P2) Speed–accuracy trade off

In the performance of many skills, the outcome is determined by both speed and accuracy. It is generally found that as the demands for accuracy increase, the speed of the movement decreases. For example, when kicking a football for accuracy, as in a penalty kick, a hard hit shot is less accurately placed than one hit less hard; in a basketball shot, the greater the distance of the throw, the less chance the ball will go into the basket; in the long jump approach, jumpers need to hit the take-off board accurately so there is a tendency to reduce their speed close to touchdown.

AN APPLICATION OF THE PHASE ANALYSIS MODEL AND MOVEMENT PRINCIPLES

A soccer penalty kick is used to illustrate how the phase analysis model and movement principles are used. The kick would typically be recorded on video and be available for repeat viewing, and inspection of individual frames. A series of still images have been extracted from such a video and presented in Figure 2.2, which cover the kick from take-off from the kicking leg on the last stride to follow-through.

Figure 2.2 Still images of a penalty kick in soccer: (a) take-off from the kicking leg, (b) last stride, (c) touchdown support leg, (d) maximum knee flexion of kicking leg, (e) contact, (f) contact rear view, (g) post-impact, (h) follow-through.

In the phase analysis template (Table 2.5), each phase is identified along with the relevant sub-phases, key moments, critical features and movement principles. The movement principles can be the abbreviated names as used earlier. It can be seen from this analysis that several movement principles may apply at the

Table 2.5 Phase analysis model template for the soccer kick.

Phase description	*Preparation (run up)*			*Retraction (of kicking leg)*	*Action (swing of kicking leg)*		*Follow-through*	
Sub-phase description	1 Place ball and withdraw	2 Approach strides	3 (fig 2.2a) Last stride	1 SL placement of (2.2b) KL hip extension (2c)	1 (fig 2.2d) KL hip forward KL knee flexes	2 (fig 2.2e) KL knee extends	1 (fig 2.2g) KL knee full extension	2 (fig 2.2h) KL knee flexes
Key moments			KL take-off	KL max hip extension (2.2c)	KL max knee flexion	Impact	KL knee fully extended	KL knee flexed
Critical features			Stride length	KL hip extension (2.2c) Opposing arm back (2.2d)	KL max knee flexion	Body posture at impact	KL knee fully extended Opposing arm forwards	KL knee flexed
Movement			ROM	ROM, SSC (stretch), AR	PDS	PDS, SSC (short-en), EPS	AR	
Principles								

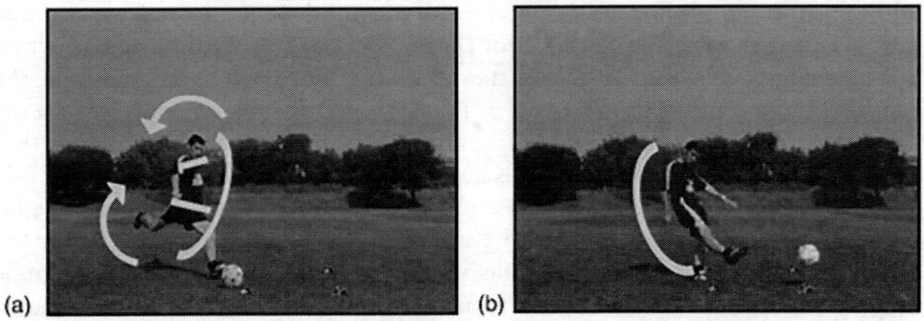

Figure 2.3 Selected images from the soccer kick with indications of important movement principles. (a) Illustration of the stretch arc (stretch shorten cycle principle), retraction of the hips and the hip-shoulder separation (both ROM – range of motion principle) and the simultaneous retraction of the kicking leg and opposite arm (AR – action reaction principle). (b) Illustration of the shorten arc (SSC – stretch shorten cycle principle) in the follow-through.

same time. Some of these movement principles can be more easily appreciated by drawing appropriate indications on the images. For example, in Figure 2.3 the stretch shorten cycle, range of motion and action-reaction principles are easily appreciated from annotations to the images. Once a phase analysis model is completed, the coach is then able to view performances of the skill with a knowledge of what to look for in the movement (critical features) and how these relate to the mechanical performance of the skill (movement principles). For example, if a player needs to improve kick speed, then the coach would reasonably look at the range of motion achieved as indicated by the length of last stride, the degree of retraction of the hips and the use of the contralateral arm. Once these aspects of technique had been improved, the coach may then focus on the coordination of the movement, specifically the proximal-to-distal sequence of the movement and the general speed of execution which would improve the effect of the stretch–shorten cycle. Once these characteristics of technique had been developed, further improvement in performance may well come from the development of muscle strength characteristics. One should note that as strength changes so too may the technique, therefore a continual monitoring of the technique used should always be made.

AN APPLICATION OF THE HAY AND REID OUTCOME MODEL

The long jump is used to illustrate how the performance outcome model can be applied. The measure of performance in the long jump is the official distance jumped. As athletes must take off in front of the foul line, they often give themselves a margin of safety by taking off a few centimetres in front of the foul line. This distance is known as the toe-to-board distance. Thus, the actual distance jumped is almost always greater than the official distance. In fact, by simple addition, the actual distance jumped = the official distance + toe-to-board distance. This is an example of rule 1 for the construction of the model. The actual distance can then be divided into the take-off distance (the distance that the centre of mass is in front of the toes at take-off), flight distance (the distance the centre of mass moves through the air) and the landing distance (the distance that the centre of mass is behind the heels at landing). This is also an example of rule 1. Of these three distances, the most important is the flight distance. In flight, the body is a projectile and so the flight distance is governed by the mechanical variables that determine projectile flight. These are the height, speed and angle of projection at release (see performance principle P1). For practical reasons, it is more convenient to combine the speed and angle together and then use the velocity components in the vertical and horizontal directions. Thus, in this example the three projectile parameters used are height, horizontal velocity and vertical velocity of the centre of mass at release. This is an example of rule 2, which is based on mechanical relationships. A full hierarchical analysis for the long jump is given in Figure 2.4 (but also see Bartlett, 2007, for a more detailed development). Note that only the main factors are followed through. Nevertheless, the model is quite comprehensive and has been used by the author to provide scientific support for international long jumpers. The model identifies clearly what needs to be measured through biomechanical analysis. For example, the horizontal velocity of the centre of mass at touchdown and take-off are key factors. These can be measured from motion analysis during competition. This provides a strong rationale for the provision of biomechanical services during competitive events. If one wants to measure factors deeper into the hierarchy, such as force, then the model implies that a force platform needs to be used. So far, this has not been done in high-level competition and it may never be done.

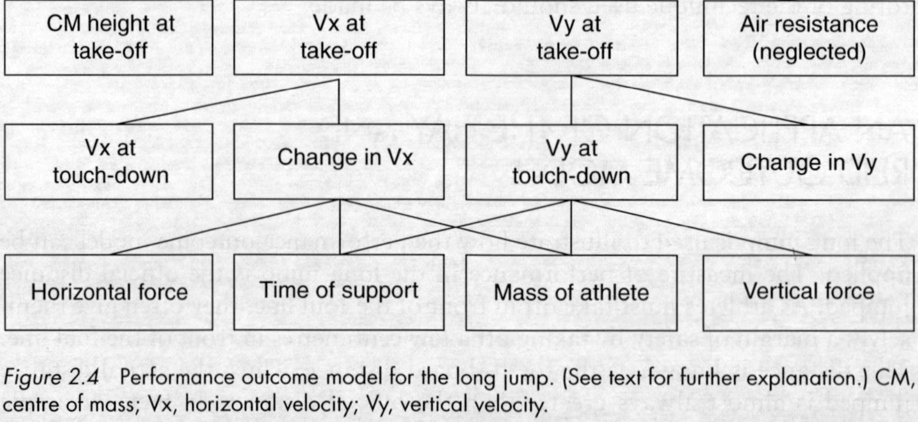

Figure 2.4 Performance outcome model for the long jump. (See text for further explanation.) CM, centre of mass; Vx, horizontal velocity; Vy, vertical velocity.

This type of information is therefore only gained in a training environment. The implication is that suitably instrumented training environments need to be available to provide comprehensive support to high-level athletes. Finally, we need to return to the issue of how this model can be used to analyse technique. Implicit in the model is the need to quantify the factors identified. This is the role of biomechanists but once this is done the results of their analyses can be used with the model to provide advice on aspects of technique. For example, it is clear in the model that performance is dependent on the horizontal velocity. Thus, it is also clear that maximizing this will be advantageous. It is also apparent from the model that vertical velocity is important, but the only place that vertical velocity can be generated is at take-off. Thus, attention should be focussed on the actions of take-off and the actions leading up to take-off. This then implies that a more detailed investigation of the technique during this phase is warranted, and that may be approached using the phase analysis model described in detail earlier.

RECENT THOUGHTS AND DEVELOPMENTS

The scope of qualitative analysis is greater than outlined earlier in the chapter. Texts on the topic (Knudson and Morrison, 2002; Bartlett, 2014) emphasise wider issues such as the preparation, observation, evaluation, diagnosis and intervention skills required to undertake a complete qualitative analysis with a purpose of improving 'performance'. This emphasises the origin of qualitative analysis which was in the coaching arena where its application was to aid the coaching process. Its formalisation, as noted earlier, is still relevant to this group, but is also appropriate to the professionalization of the sports biomechanists whose services are now sought by national governing bodies of sport in many countries. The reader is encouraged to consult the Knudson and Bartlett texts for a wider appreciation of the method.

While texts have focussed on the wider issues, researchers have attempted to solve other problems. One in particular which is worth noting is the concept of 'technical level'. Any skill may be performed badly or well and as beginners learn they tend to move from a poor to a good demonstration of the skill. The research issue is whether this progress can be determined using qualitative analysis. Marques-Bruna et al. (2007) attempted to determine the technical proficiency of 187 boys and girls aged 6–11 years, and 31 male and female adults in a stationary ball-kicking task. Phase analysis was used to divide the event into five relevant components, and for each, three levels of technical proficiency were defined and scored. For example the first component was the approach and was classified as straight (1 point), angled (2 points) and curved (3 points). The other components were opposite arm movement, foot placement, contact pattern and follow-through. The highest technical level was associated with a mature skilled kick and the greater technical proficiency was associated with a higher score. The results were clearly able to distinguish between ages and sexes, with older children demonstrating a higher technical proficiency than younger children, and boys showing a higher technical proficiency than girls. These differences were evident in each component as well as overall. Interestingly the adults were little better than the 11-year-old boys. This novel approach introduced a new concept into the literature and has shown it to be sensitive to skill development of children on a large scale. Such an analysis using a quantitative approach would be impossible.

Finally it is worth commenting on the advancement of computer technology. The introduction of low cost video analysis has enabled qualitative analysis to occur, but this is now several decades old. The introduction of low cost computing has enabled quantitative analysis to progress. So the question is whether low cost computing can be combined with image analysis to advance qualitative analysis? Certainly developments have taken place which allow drawings and other information to easily be superimposed onto images to produce diagram-pictures, as in Figure 2.3, which could easily be used in the field in real time to provide feedback to performers. Computing technology is advancing rapidly so perhaps the best advice is to be . . . observant!

CONCLUSION

This chapter has sought to introduce the methods relevant to the qualitative analysis of sports (and exercise) skills. Qualitative analysis is appropriate to the biomechanical analysis of technique and two major approaches have been identified for this. The phase analysis model is appealing in that it is based on a sequential breakdown of performance, performed from visual images gained, typically, from video analysis. By identifying phases and sub-phases of movement, the key moments and critical features, it is possible to identify the principles that govern performance. Using these principles, faults can be diagnosed from which it is hoped that performance may be improved. In order to use this model effectively, a sound knowledge of movement principles is needed. These are not readily available in the literature so the opportunity has been taken to outline them in this chapter. In contrast, the second approach is based on performance outcome. This leads to a hierarchical model of performance which identifies the relationships and mechanical factors that govern performance. It is implicit in this model that many of these factors need to be measured quantitatively. Based on this information it is possible to gain an insight into the key factors that influence performance and attend to these directly (such as the importance of approach speed in the long jump) or to use this information to guide a more 'technique' orientated analysis using the previous phase analysis model. The tools provided in this chapter should enable the enthusiastic student of sport and exercise science to undertake an effective analysis of technique using qualitative methods.

REFERENCES

Bartlett, R. (2007) *Introduction to Sports Biomechanics*, 2nd edn. London: Routledge.

Bartlett, R. M. (2014) *Introduction to Sports Biomechanics: Analysing Human Movement Patterns*, 3rd edn. London: Routledge.

Chow, J. W. and Knudson, D. V. (2011) 'Use of deterministic models in sports and exercise biomechanics research', *Sports Biomechanics*, 10: 219–233.

Hay, J. G. and Reid, G. (1982) *Anatomy, Mechanics and Human Motion*. Englewood Cliffs, NJ: Prentice-Hall.

Knudson, D. V. (2002) 'Qualitative biomechanical principles for application to coaching', *Sports Biomechanics*, 6: 109–118.

Knudson, D. V. and Morrison, C. S. (2002) *Qualitative Analysis of Human Movement*, 2nd edn. Champaign, IL: Human Kinetics.

Lees, A. (2002) 'Technique analysis in sports: a critical review', *Journal of Sports Sciences*, 20: 813–828.

Marqués-Bruna, P., Lees, A. and Grimshaw, P. (2007) 'Development of technique in soccer', *International Journal of Coaching Science*, 1(2): 51–62.

CHAPTER 3

ASSESSING MOVEMENT COORDINATION

Peter F. Lamb and Roger M. Bartlett

INTRODUCTION

One criticism of sports biomechanists has been that we focus too much on discrete data, such as jump take-off angles, or durations of movement phases, or maximum joint torque in a throw, thereby discarding much of the richness of information contained in time-series data. This becomes increasingly the case when we consider movement coordination, something important also to motor learning or motor control specialists. The interpretation of various coordination patterns is the topic of this chapter. These coordination patterns are particularly relevant when we ask the question 'How are sports and exercise movements coordinated to produce the desired outcome?'

To obtain the coordination patterns that are the focus of this chapter, we need to be able to measure joint angles, even though the analysis of these patterns can be almost entirely qualitative, as well as quantitative. Two-dimensional (planar) joint angles can be easily measured using software packages for qualitative biomechanical analysis, such as Kinovea, or, usually more accurately, using quantitative video analysis packages, such as SIMI Motion (see also Chapter 4). The focus in most of this chapter will be on data obtained in the sagittal plane. For analysis of movement patterns that occur in multiple planes, three-dimensional motion analysis, using video or on-line analysis systems, is usually necessary (see also Chapters 4 and 5). The chapter will also emphasise the qualitative and semi-quantitative analysis of the coordination patterns that we will discuss, mainly because we consider movement coordination to be a 'qualitative' process. We feel strongly that even quantitative analysts need to be able to describe and to analyse coordination patterns qualitatively, if they are to understand them fully.

Coordination patterns can add substantially to our understanding of sports and exercise movements and help us to assess ways in which changes might be made to improve performance. Because different individuals find unique solutions to the demands of a sports or exercise task, under the various constraints of that task, the environment and their own organism, intra-individual studies are usually more productive than inter-individual ones in shedding light on how movements might be better coordinated. We might, for example, compare different speeds of locomotion, or different stroke rates in rowing, or track an athlete recovering from injury.

Before looking at how we interpret patterns of coordination, let us begin by considering what we mean by this important term 'coordination'. One of the generally accepted universal principles of movement is, 'Mastering the many degrees of freedom involved in a movement' (Bartlett, 2014, p. 96). This is one statement of what movement coordination involves. A rather longer definition, which elaborates on the one in the previous sentence, introduces the idea of 'coordinative structures'. This viewpoint sees the acquisition of coordination as constraining the degrees of freedom into coordinative structures, which are functional relationships between important anatomical parts of a performer's body, to perform a specific activity. An example would be groups of muscles or joints temporarily functioning as coherent units to achieve a specific goal, such as hitting or catching a ball. As muscles act around joints, this explanation leads us to look at joints and their interrelationships to gain an initial insight into how sports and exercise movements are coordinated.

In the following subsections we will look at the qualitative analysis of several different coordination patterns. Angle–angle diagrams are graphs of one joint angle as a function of another. The focus is on how one angle changes with changes in a second angle; in other words we focus on how the two angles 'co-vary' rather than how they each evolve with time. Angle–angle diagrams have been used extensively in the study of movement coordination, particularly in locomotion (e.g. Hershler and Milner, 1980; Kutilek and Farkisova, 2011). They can also be three dimensional although these are far more difficult to interpret. Cross-correlation functions look at how the correlation coefficient between two time series, such as joint angles, changes as one time series lags behind the other. As with the other correlation patterns, these are normally 'pairwise' – each graph looks at how two joints are coordinated.

Phase planes, as used in studying human movement coordination, are normally graphs of the angular velocity of one joint as a function of the angle of that same joint. The focus is on the so called coordination dynamics of that joint. Phase planes are used extensively in the study of movement coordination; usually the phase of one joint relative to another is of most interest to the analyst (see Continuous Relative Phase section later in this chapter). More recently, the use of self-organizing maps has enabled the visualisation and study of multi-dimensional coordination – that of multiple joints, which is what most sports and exercise movements involve.

All of these coordination patterns are discussed in detail in the following sections. The focus is on interpreting these graphs qualitatively and semi-quantitatively; developing such interpretive skills is important to appreciate how coordination patterns can then be used to try to improve sport and exercise performance.

ANGLE–ANGLE DIAGRAMS

Time series involving several angles can be difficult to interpret for coordination. An alternative is to plot angles against each other – these are called angle–angle diagrams; we could plot three angles in this way to form a three-dimensional plot, but this is rarely done.

Several forms of coordination can be brought to light through angle–angle diagrams. Some of these relate to the differences between synchronous and sequential coordination in, for example, jumping and throwing (Bartlett, 2014). In synchronous – or simultaneous – coordination, joints move more or less together. These synchronous movements may be 'in-phase' coordination, as when the hips and knees flex during the downward phase of a standing vertical jump (SVJ) and extend during the upward phase (this is shown in Figure 3.1(a), which is similar in many respects to in-phase turning point coordination in Figure 3.1(b), except that the SVJ is a discrete movement so has a turning point coordination change only at the bottom of the lowering phase). Another example is the synchronous extension of the hips and the knees during the drive phase of rowing (see Figure 3.4). If the two angles change at the same rate, a linear relationship, such as that of Figure 3.1(a), would result; this is rarely seen in human movements in sport. More often, the joints will show in-phase 'turning point' coordination since their angles change at different rates, as in Figures 3.1(b) and (e). The synchronous movements may also be out-of-phase. This occurs when one joint flexes and another extends in the sagittal plane, or when one joint abducts while another adducts in the frontal plane: we then have 'anti-phase' or 'out-of-phase' coordination. An example is the action phase of a darts throw in which the upper arm abducts while the wrist adducts from a partially abducted position at the start of the phase (the elbow extends at the same time, also in synchrony but in a different plane of motion). Linear 'anti-phase' coordination is shown in Figure 3.1(c) (again, this is rarely found in human movements in sports) and anti-phase turning-point coordination in Figure 3.1(d).

Figure 3.1(f) shows 'phase offset' or decoupled coordination. In this type of coordination, the two joints change their relative roles during a cycle of a movement, switching between in-phase and anti-phase coordination. Reading from point A anticlockwise, both angles flex in-phase, until B, then Angle 2 continues to flex while Angle 1 extends (anti-phase). From points C to D, both angles extend in-phase and, finally, Angle 1 flexes from points D to A while Angle 2 continues to extend (anti-phase). Finally, we have sequential coordination, to which we will return later in this section. Coordination in actual human movements is often more complex than these basic patterns, as illustrated by the hip–knee angle–angle diagram for one running stride in Figure 3.2. Please note that the diagram has been slightly 'massaged' so that it forms a continuous loop, which is rarely the case owing to movement variability and measurement errors.

Reading around Figure 3.2(a), starting at the lower right hand spike at 'a', which corresponds roughly to touchdown, or heel strike, and progressing anticlockwise, the pattern is as follows. At the start of the stance phase, the hip and knee both flex until 'b'; then, briefly, the knee continues to flex while the hip extends to 'c'. From 'c' to 'd' the two joints extend in-phase. From 'd', which is

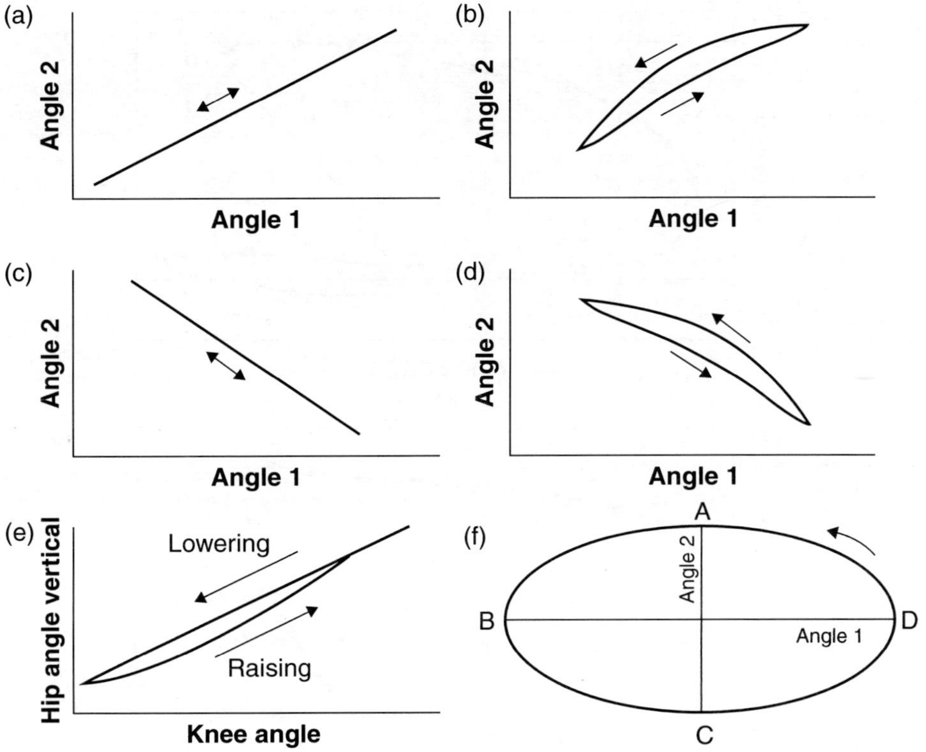

Figure 3.1 Basic types of coordination. In-phase: top left (a) linear; top right (b) turning-point coordination. Anti-phase: middle left (c) linear; middle right (d) turning-point coordination. Bottom left: hip and knee angle coordination in a standing vertical jump with countermovement (e) – compare this with (b). Bottom right: (f) phase offset or decoupled coordination.

roughly at toe-off, another brief period until 'e' sees the knee flex while the hip extends at the start of the swing phase. From 'e' until 'f' both joints flex in-phase during the next part of the swing phase, after which the knee extends while the hip continues to flex until 'g'; both joints then extend in-phase until around touchdown. It is instructive to repeat this description of joint movements for the same running stride but looking at the ankle–knee and ankle–hip joint couplings, in Figures 3.2(b) and (c).

Angle–angle diagrams have both advantages and disadvantages. Their advantages include that we do not have to flip between angle–time graphs to see how angles co-vary, and that we can pair joint angles of interest easily to show how they co-vary. These graphs show coordination patterns qualitatively, which can facilitate comparisons, for example, between individuals and for one individual during rehabilitation from an injury. They also show all of the fine details of how the two joints change coordination during the movement, as in Figure 3.2(a). We can also compare patterns between, for example, running and walking. Many attempts have been made to quantify angle–angle diagrams, such as vector coding (e.g. Glazier and Wheat, 2006); such a reduction of a rich qualitative pattern to a few numbers seems unnecessary to us and ignores the

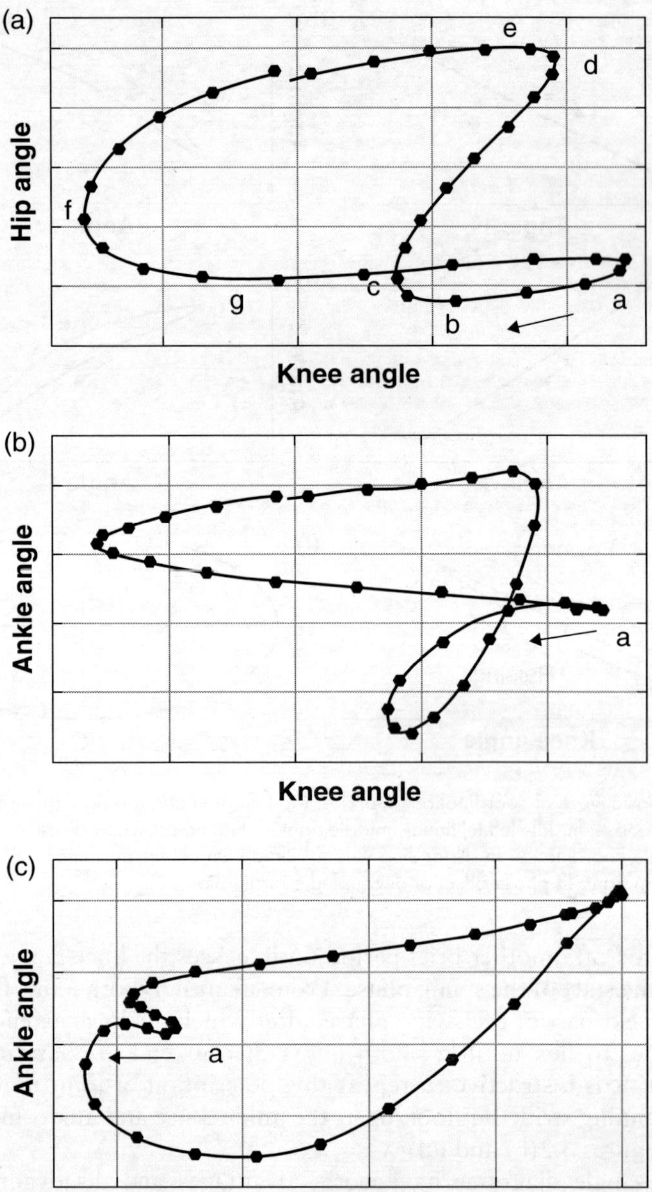

Figure 3.2 Angle–angle diagrams for one 'ideal' treadmill running stride: top (a) hip–knee coupling; middle (b) ankle–knee coupling; bottom (c) ankle–hip coupling.

saying 'a picture is worth a thousand words'. Few attempts have been made to distinguish patterns qualitatively; one of the very few is known as 'topological equivalence'. Two shapes are topologically equivalent if one can just be stretched – albeit by different amounts in different places – to form the other; two shapes are not topologically equivalent if one has to be 'folded' rather than just stretched to form the other (see, e.g. Bartlett and Bussey, 2012). Simplistically, this means that if the shapes have different numbers of loops, they are not topologically equivalent; they are then qualitatively rather than just quantitatively different, as for the ankle–knee, hip–knee and ankle–hip couplings when comparing the running angle–angle diagrams in Figure 3.2 with those for walking in Figure 3.3.

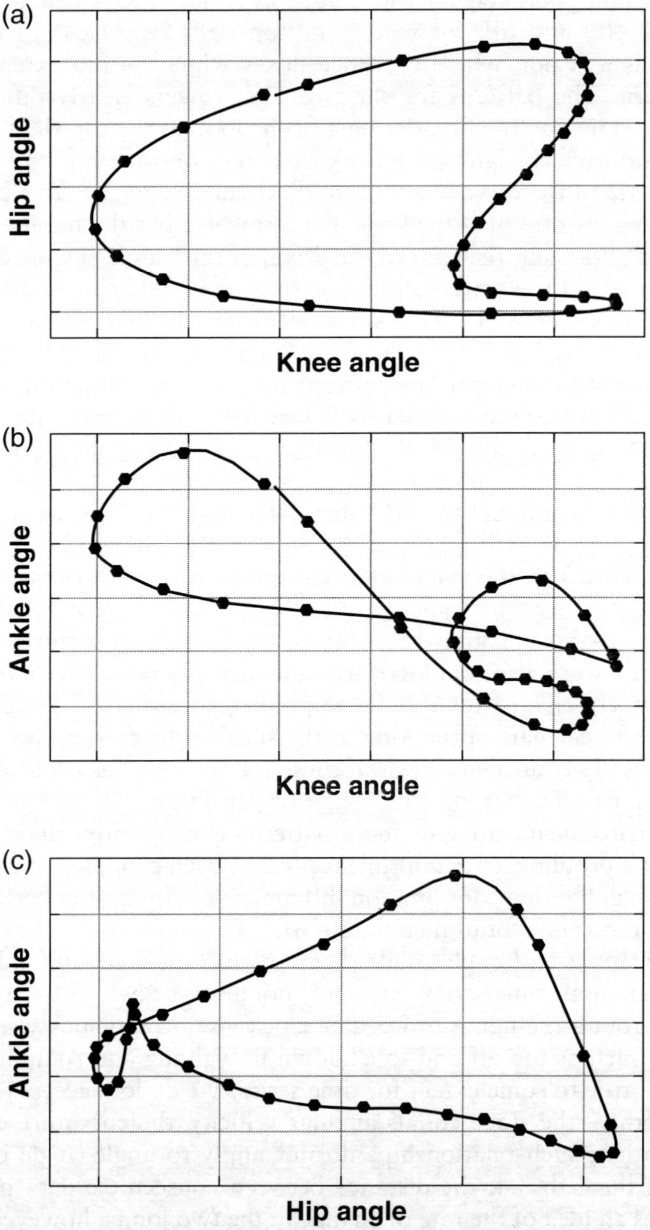

Figure 3.3 Angle–angle diagrams for one treadmill walking stride: top (a) hip–knee coupling; middle (b) ankle–knee coupling; bottom (c) ankle–hip coupling.

An important point to make here, and it applies to other coordination patterns too, is that of inter-individual differences. Consider, for example, Figure 3.4, which shows hip–knee angle–angle diagrams for four rowers performing a race trial on a rowing ergometer. Figures 3.4(a) and (b) are for two club standard rowers and Figures 3.4(c) and (d) are for two high performance rowers. It should be perfectly clear that there is no 'club' or 'elite' template here. All four rowers show a coordination pattern that is close to synchronised in-phase turning-point coordination; but this is not strictly true for the rowers in Figures 3.4(a) and (d), for whom the top right loop close to the finish of the drive has a region where the knee flexes while the hip extends, an anti-phase pattern. The patterns for the two club rowers clearly differ, with one (club rower 1) having the broader part of the loop at the top right of the curve (Figure 3.4(a)), which signifies faster knee extension than hip until max extension at the end of the drive when both joints start flexing; at this point the hip flexes at a greater rate than the knee. Club rower 2 has the broader part of the loop at the bottom left, the start of the drive, or the 'catch' (Figure 3.4(b)). Both have two loops in their pattern, although these may not be immediately obvious from the figures; Figure 3.4(a) has a small second loop at the bottom left and Figure 3.4(b) at the top right of the angle–angle pattern. They are topologically equivalent but still different. The patterns for the two high performance rowers also differ. The rower represented by Figure 3.4(c) (high performance rower 1) has a single loop pattern, with the loop being broader at the start of the drive, whereas the rower in Figure 3.4(d) (high performance rower 2) has the broader part of the loop at the finish of the drive. The two patterns are also not topologically equivalent, with high performance rower 2 having a clear two-loop pattern in contrast to the single-loop pattern of high performance rower 1. If we compare across performance standards, we note the general shapes of the hip–knee angle–angle diagrams for club rower 1 and high performance rower 2 are similar; they are also topologically equivalent, as both have two loops. The general shapes of club rower 1 and high performance rower 2 are also similar in having the broader part of the loop at the start of the drive; however, they are not topologically equivalent. Each of these four rowers has evolved a hip–knee coordination pattern that 'matches' their organismic constraints to those of the task and environment; none of these patterns is inherently 'right' or 'wrong'. Coordination profiling, over multiple trials, has been proposed as a comprehensive way of highlighting coordination differences or similarities between individuals (see, for example, Button et al., 2006).

Disadvantages of angle–angle diagrams include some unfamiliarity compared to joint angle time series. Also, it is not always obvious from the diagram which way round the figure proceeds – clockwise or anticlockwise – or where key events, such as toe-off and touchdown in walking and running, occur; the latter is also true to some extent for time series. We do lose access to time-series shape patterns – the slope equals angular velocity, the curvature equals angular acceleration; such relationships do not apply to angle–angle diagrams. In Figures 3.2 through 3.4, the distance between consecutive data points (black circles) gives an idea of the rate of change in the two joints, however, it is a little more difficult to interpret since it is the combined rate of change of two joint

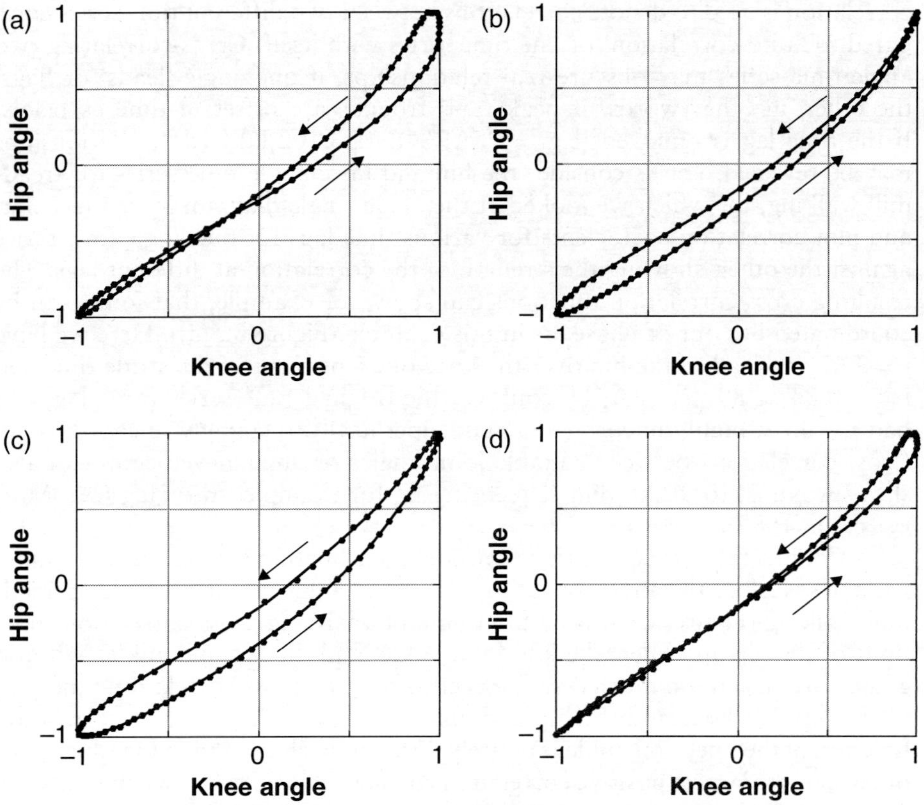

Figure 3.4 Hip–knee angle–angle diagrams for four rowers early in a five-minute race trial on a rowing ergometer, one 'stroke' only shown on each: top left (a) and top right (b) club standard rowers; bottom left (c) and bottom right (d) high performance rowers. The 'catch' is at the bottom left of each angle–angle diagram and the 'finish' near the top right. Similarly, moving from bottom left to top right on the diagram indicates the drive phase and top right to bottom left, the recovery phase. The angles are all normalised to the range – 1 to +1.

angles. To obtain velocity and acceleration information for each joint, we need to study the time-series data from which our angle–angle diagram was plotted.

CROSS-CORRELATION FUNCTIONS

Here, we will consider another graphical representation of coordination between joints – cross-correlation functions – and look at the strengths and weaknesses of this approach. Cross-correlations are similar to Pearson product moment correlations, but involve correlating variables (often angles) from two time series. They can be easily implemented by the function PEARSON in Excel; e.g. = PEARSON(D2:102,E2:E102), where the data for one angle are in cells D2 to D102 and for the second angle in cells E2 to E102. The term 'cross' correlation is used to denote correlations between two different time series compared to 'auto-correlation' of one time series with itself. Cross-correlating two angle time series may obscure real relationships, if one angle 'leads' or 'lags' the other, in other words, if peaks and troughs are offset in time as below. If the time lag is removed (e.g. = PEARSON(D3:102,E2:E101), relationships may be revealed. Let us consider the hip and knee angle time series for treadmill walking, for which we looked at the angle–angle diagram in Figure 3.3(a) and plot correlation coefficients for various time lags for one angle time series against the other, showing the strength of the correlation at different lags. The resulting cross-correlation functions can show, for example, that joints can be coordinated but out of phase, as in this example (Figure 3.5(a)). Here we have $r = 0.67$ at a lead of the hip over the knee of 25 per cent of the stride time and $r = -0.84$ at 80 per cent; by contrast, the original time series (time lag = 0) had $r \sim 0.02$, highly uncorrelated and, superficially, seemingly uncoordinated. Cross-correlations between suitable joint angles or angular velocities can also show proximal-to-distal joint sequencing in, for example, throwing (e.g. Morriss et al., 1997).

Conjugate cross-correlations consider coordination between three or more variables, such as the angles of the hip, knee and ankle in our example of treadmill walking (Figure 3.5(b)), by plotting and analysing cross-correlation functions of the hip and knee, the knee and ankle, and the hip and ankle. Without going in to this in too much detail, it is clear that the cross-correlation functions of the hip and knee and the hip and ankle are inverted. The cross-correlation function of the knee and ankle is entirely different in shape and in the time leads, or lags, to the largest positive correlation coefficients. Rules for which joints can be meaningfully correlated when using conjugate cross-correlations were proposed by Amblard et al. (1994); a simpler rule is only to seek to correlate joint angles or angular velocities for joints that are clearly correlated in a particular movement.

Cross-correlation functions can reveal aspects of coordination not apparent in the other approaches that we have considered in this chapter; for example, they show whether one joint lags behind another. This is particularly useful in studying proximal-to-distal sequential coordination, as in most throws for distance; the comparisons need to be pairwise (two joints at a time), a limitation of this approach. Although the calculation of cross-correlation functions might seem to be 'quantitative', the interpretation of them here is essentially qualitative, or semi-quantitative if we include the time lags and correlation coefficients. The cross-correlation function patterns are easy to interpret, although some people find the underlying concepts somewhat difficult.

Sadly, in our view, cross-correlation functions have become 'unfashionable' in sports biomechanics, probably because of mistaken beliefs about what statistical assumptions need to be satisfied in their use. If we are only interested in calculating correlations, and establishing at which lag the maximum or minimum occurs, then the only underlying assumption is that the two variables are linearly related. It should be obvious that there must be a meaningful relationship between the two variables to be correlated, otherwise erroneous results will be obtained. Human movement dynamics are not, however, normally linear, but this is not an insurmountable difficulty as we can, for example, logarithmically

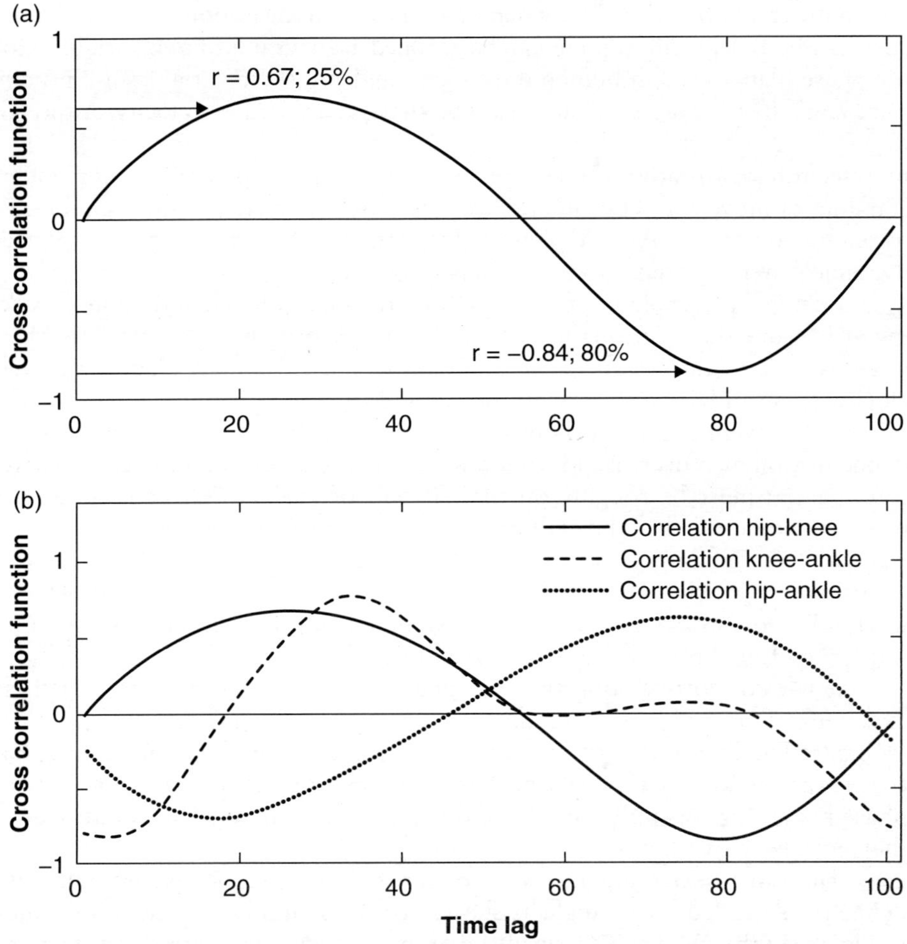

Figure 3.5 (a) Cross-correlation function for the hip and knee angles in treadmill walking for a specific individual at preferred walking speed. (b) Conjugate cross correlation functions between the hip and knee, knee and ankle, and hip and ankle angles in treadmill walking for a specific individual at preferred walking speed.

transform the data. If we wish to test hypotheses, for example about the statistical significance of the correlation coefficient, the same assumptions must be met as for Pearson product moment correlations; these are clearly described by, for example, Howell (2010).

PHASE PLANES

Perhaps the main criticism of angle–angle diagrams is that while they show how coordination between the two joints changes (by detailed analysis as described earlier) they do not tell us why these changes take place. To try to do that, we need a different approach. Phase planes are based on the notion that any system, such as a body segment or joint, can be graphed as diagrams of two variables; for the phase planes used in human movement analysis, these variables are usually joint angle (sometimes segment absolute angle) and angular velocity. Although the relevance of a phase plane for a single joint to coordination between joints may seem hard to fathom, phase planes turn out to be pivotal for our understanding of movement coordination, as they are one method which enables us to obtain continuous relative phase (next section). They are an integral part of dynamical systems theory (see, e.g. Kelso, 1995).

Example phase planes, for the hip and the knee joints in a treadmill walking stride, are shown in Figure 3.6(a). Their description – although not their analysis – is straightforward. First, let us ask whether the phase plane progresses clockwise or anticlockwise with time around its loop – it must do one or the other. By convention, in sports biomechanics for two-dimensional analysis, we define flexion as a decrease in joint angle and extension as an increase. Therefore, flexion must be roughly on the left half of Figure 3.6(a) and extension towards the right half – this will partly depend on the normalisation used (discussed later). Similarly, a flexion velocity is negative and an extension velocity is positive; so, flexion must be below the horizontal axis (zero angular velocity line) in Figure 3.6(a) and extension above it. The phase planes of Figure 3.6(a) must, therefore, progress clockwise with time.

It is also worth noting that the phase planes for the hip angle in these examples of treadmill walking and running are not topologically equivalent (compare Figure 3.6(a) with (b), respectively); the phase plane for running has two loops while that for walking has only one. However, the example knee phase planes for treadmill running and walking are topologically equivalent as they both have two loops.

The value of phase planes should start to become more evident when we define the so-called phase angle as shown for the example of treadmill running in Figure 3.6(b). We have changed the graph so that it is 'centred' on its mean value, for reasons that we discuss briefly below. The phase angle, φ, is calculated at time t_i as

$$\varphi(t_i) = \arctan\left(\frac{\omega(t_i)}{\theta(t_i)}\right)$$

where ω and θ represent angular velocity and angular displacement, respectively.

If we now subtract the phase angle – defined anticlockwise from the right horizontal – for one joint from that for a second joint, at the same instant, we define a variable known as relative phase. Here, we subtract the knee phase angle from that for the hip, (this is the usual way of doing such things – subtracting the proximal joint phase angle from that for the distal joint). We can do this for every data point (or time instant) in the phase plane (here defined by the running cycle) to arrive at values of this relative phase as a function of time, which can then be graphed and is known as 'continuous relative phase'.

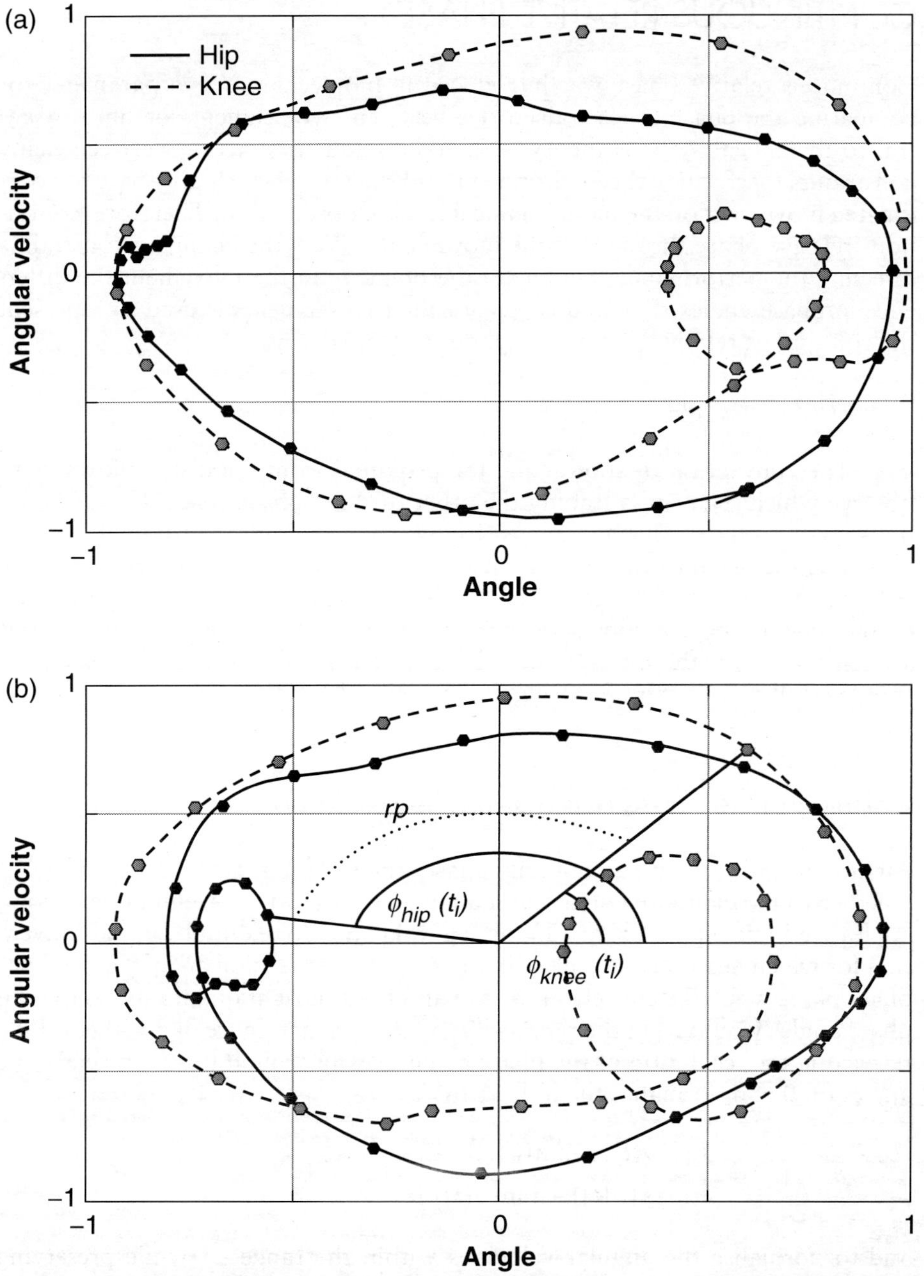

Figure 3.6 (a) Centred and normalised phase planes for the hip and knee for one treadmill walking stride. (b) Superimposed phase planes for the hip (continuous curve) and knee (dashed curve) joints in one treadmill running stride plus definition of their phase angles ($\phi_{hip}(t_i)$ and $\phi_{knee}(t_i)$) and the relative phase angle (*rp*).

CONTINUOUS RELATIVE PHASE

Continuous relative phase was introduced in biomechanics as a parameter for comparing the phase space trajectories between two segments or limbs. Central to dynamical systems theory is the continuous interaction between many constraints (performer, the environment and the task), which gives rise to coordinated movement on the biomechanical level of observation. Therefore, continuous relative phase throughout the movement reflects the changing constraints affecting the performance. For each time point, t_i, in the movement the difference in phase angles φ_{distal} and $\varphi_{proximal}$ of the two segments is used to represent the continuous relative phase throughout the movement.

$$crp(t_i) = \varphi_{distal}(t_i) - \varphi_{proximal}(t_i)$$

The convention of subtracting the proximal from the distal allows us to observe which segment is leading the other through phase space. If the continuous relative phase value at t_i is positive, we can say that the proximal segment is leading the distal through phase space, and vice versa (Stergiou et al., 2001). Because there is some disagreement between researchers on when the equation on the previous page is considered appropriate for calculating phase angles, this section will cover the most commonly reported method, using the phase portrait, as well as a method for using analytic signals.

Calculating the phase angle using the phase portrait

An important issue in calculating phase angles using the phase portrait is whether to normalise the angles and angular velocity data. Definitions of joint angles are somewhat arbitrary. In the sagittal plane, for example, in the convention for which flexion is a decrease in the angle and extension an increase, a knee phase plane would lie entirely in the top and bottom right quadrants, so that all phase angles would be in the range +90 to −90° (in fact, more like +50 to −40°), restricting the resolution of the phase angle. A way around this is to centre the angles on 0 and normalise them to ±1 by

$$\theta_{norm}(t_i) = 2\left(\frac{\theta(t_i) - \min(\theta(t))}{\max(\theta(t)) - \min(\theta(t))}\right) - 1$$

and to normalise the angular velocities within the range ±1 while preserving zero by

$$\omega_{norm}(t_i) = \frac{\omega(t_i)}{\max(|\omega(t)|)},$$

which is meaningful (e.g. Miller et al., 2010). This then gives a range of phase angles from 0 to 359°, improving their resolution (as in Figure 3.6). We recommend to centre and normalise phase planes, but there is some disagreement on this (compare, for example, Hamill et al., 2000; Kurz and Stergiou, 2002; Lamb and Stöckl, 2014; Peters et al., 2003).

Although normalising the phase angles is preferred, there are other issues that can arise from normalisation. First, Figure 3.6 shows a loop in the phase space trajectory for the knee, which traverses the right horizontal. Since the goal of normalisation is to obtain a roughly circular phase portrait, large loops are considered by some to introduce frequency artefacts (see Fuchs et al., 1996). However, with a relatively consistent movement like the gait cycle, the results of continuous relative phase using phase angles from the phase portrait have been argued to be reliable (Peters et al., 2003). A second issue involves selecting the normalisation factors. For example, when comparing multiple trials, if the maximum knee angle in trial 1 is used to normalise trial 1 and the maximum knee angle for trial 2 is used to normalise trial 2, then the phase portraits for each respective trial have been scaled by different factors, thus affecting their phase angles differently. Again, this does not pose much of a problem if the movement being examined is relatively consistent between trials; however, there are no rules for deciding whether the consistency between trials is satisfactory. One way to avoid this problem is to use the maximum from a group of trials as the scaling factor for all trials within the group, as in Figure 3.6 (e.g. Hamill et al., 2000; Trezise et al., 2011); this method, however, is susceptible to outliers.

Calculating the phase angle using analytic signals

A different method for calculating the phase angle, which avoids the previously mentioned limitations, is based on the measured joint angle, $\theta(t)$, and its Hilbert transform, $H(t)$. After first centring the joint angles around zero by

$$\theta_{centered}(t_i) = \theta(t_i) - \min(\theta(t)) - (\max(\theta(t)) - \min(\theta(t)))/2$$

the Hilbert transform can be calculated. The Hilbert transform creates, from the measured joint angle, a complex analytic signal, $\zeta(t)$ defined as

$$\zeta(t) = \theta_{centered}(t) + iH(t)$$

Based on the complex signal, the phase angle at time t_i can be calculated by

$$\varphi(t_i) = \arctan\left(\frac{H(t_i)}{\theta_{centered}(t_i)}\right)$$

With the phase angle calculated using the analytic signal, continuous relative phase can be calculated the same as above for the phase portrait method. The result is a signal representing the phase difference between two joints as a function of time and is free of frequency artefacts (for more details, see Lamb and Stöckl, 2014). Calculating the phase angles using analytic signals is more computationally intensive; however, built-in functions for calculating the Hilbert transform come with analysis software such as MATLAB.

Figure 3.7(a) shows the time-series curves for three strides of treadmill running for the knee and hip joint angles in the sagittal plane. The phase portrait for the same variables is shown in Figure 3.7(b); here the joint angles and their angular velocities have been normalised as mentioned earlier. In Figure 3.7(c), continuous relative phase for the three strides is shown, as the legend indicates, the dotted line represents continuous relative phase calculated from phase angles using the phase portrait method and the solid line from phase angles using the Hilbert transform method. Comparing both methods in Figure 3.7(c) clearly shows that the relative phase between the knee and hip coupling is most in-phase immediately after toe-off and most out-of-phase shortly after touchdown. There is some variation between the phase portrait method and the Hilbert transform method; we suggest the reader carefully consider the movement being analysed and whether the phase space trajectories are consistent enough to use the phase portrait method.

Regardless of the method used to calculate the phase angle, there are still several limitations associated with using continuous relative phase. First, only the coordination between two entities – normally two joints in human movement – can be studied and, as we have already mentioned, most human movements involve far more than two joints. Second, is the possible need for the use of circular statistics to analyse continuous relative phase data. The use of continuous relative phase assumes, for any statistical tests, that the datasets are sinusoidal – or very nearly so (this is mostly fairly true for cyclic movements) and 'stationary' – that is, the statistical description of the dataset is invariant with time, which will not be true, for example, if impacts are involved. The next

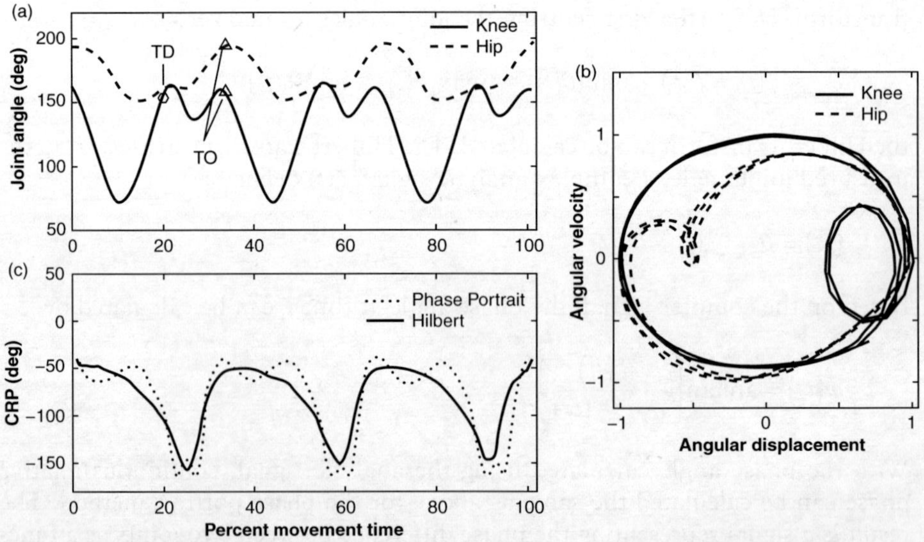

Figure 3.7 (a) Knee and hip joint angles as a function of time for treadmill running; 'TD' indicates touchdown and 'TO' indicates toe-off for the first stride. (b) Normalised phase portrait for the variables in (a). (c) Continuous relative phase for the variables in (a) calculated using phase angles from the phase portrait method (dashed line) and the Hilbert transform method (solid line).

section deals with a more comprehensive method for assessing coordination, which we argue deals well with both the limitations of continuous relative phase identified here.

SELF-ORGANISING MAPS

Sports and exercise movements usually involve the coordination of many segments and the muscles that operate them. Unfortunately, the methods for studying coordination, so far discussed, limit us to the coupling between two joints for angle–angle diagrams, two joint angles or two joint angular velocities for cross-correlation functions, and two joint angles and their angular velocities for phase planes and continuous relative phase. In order to incorporate more of the segments involved in a particular movement in our assessment of coordination, we look to artificial neural networks (ANNs).

ANNs represent an attractive method for analysing sports and exercise techniques mainly because of their non-linear properties and their characteristic ability to discover patterns in data. Self-organising maps (SOMs) are a particular type of ANN particularly useful for visualising and clustering high-dimensional data (Kohonen, 2001). For sport and exercise biomechanists, the use of SOMs comes in their ability to map the high-dimensional coordination of the original movement to a low-dimensional, visualisable output.

The output of a SOM is commonly visualised as a layer of grid nodes, each with an associated weight vector, which is connected to a layer of input nodes (Figure 3.8). The dimensionality of the nodes' weight vectors is the same as the dimensionality of the input vectors. For example, if the action we are studying is a running stride and the variables representing the running stride are (left and right): ankle plantar-dorsiflexion, knee flexion-extension, hip flexion-extension, hip adduction-abduction and hip internal-external rotation, then we have ten-dimensional input vectors and thus also ten-dimensional weight vectors in the output. Notice that in Figure 3.8 an input vector, or a node in the input layer, represents a coordination state at time t_i, so a movement time-normalised to 101 time frames will be represented by 101 input vectors. The representation of one time frame with one input vector is most commonly done in sports biomechanics to map the changing coordination throughout the movement (e.g. Bauer and Schöllhorn, 1997; Lamb et al., 2014). Before beginning a SOM analysis one must consider which biomechanical variables will be used for training. Again, there are no universal rules for which variables to include, but as general advice:

- The variables should be relevant to the movement;
- Although SOMs deal well with redundancy, avoiding obviously redundant variables will simplify the interpretation;
- We encourage the reader to use 3D joint angles as more subtle changes in multi-dimensional coordination may give new insight to those gained by pairwise coordination assessment (e.g. continuous relative phase);
- EMG and kinetics have been used less commonly in SOM analysis; experimentation with these variables may be informative.

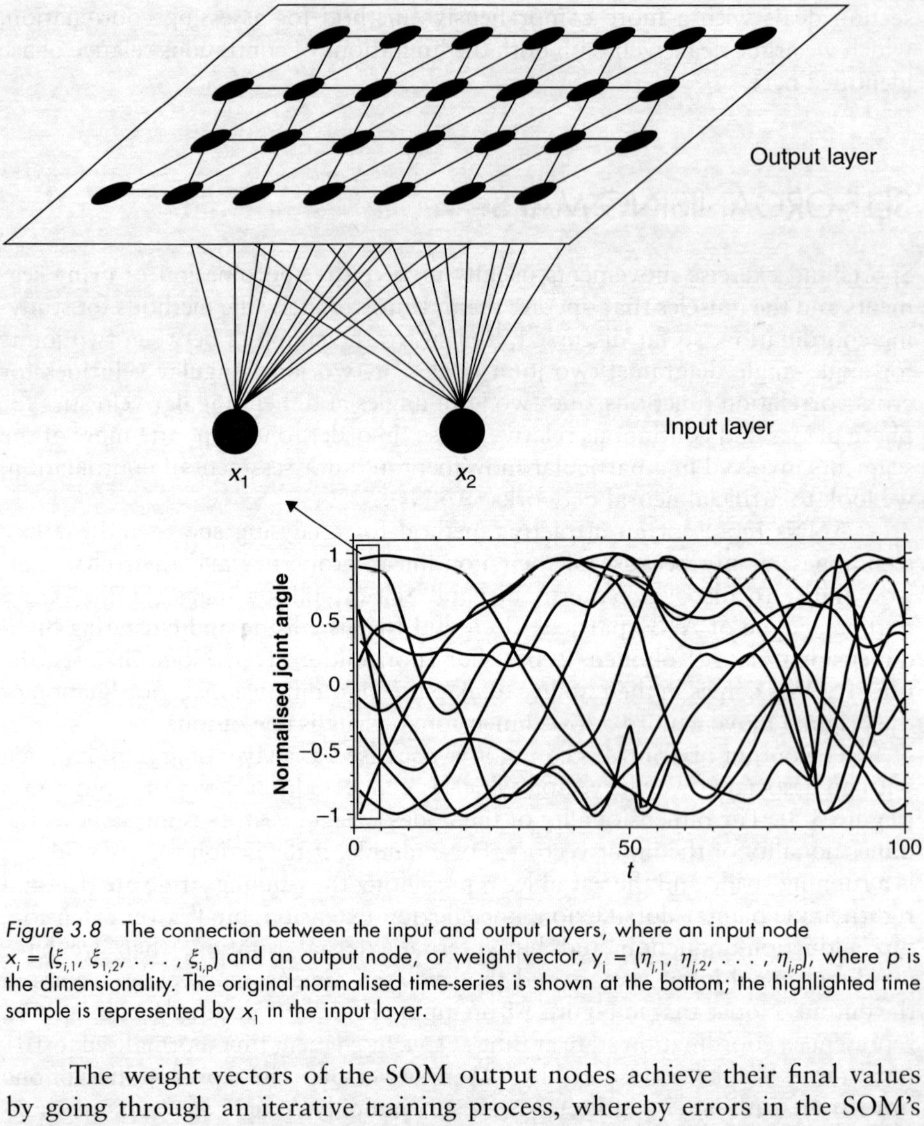

Figure 3.8 The connection between the input and output layers, where an input node $x_i = (\xi_{i,1}, \xi_{1,2}, \ldots, \xi_{i,p})$ and an output node, or weight vector, $y_j = (\eta_{j,1}, \eta_{j,2}, \ldots, \eta_{j,p})$, where p is the dimensionality. The original normalised time-series is shown at the bottom; the highlighted time sample is represented by x_1 in the input layer.

The weight vectors of the SOM output nodes achieve their final values by going through an iterative training process, whereby errors in the SOM's representation of the input distribution are decreased each iteration until the weight vectors' values eventually stabilise. To train a SOM the following steps are necessary:

1 *Normalisation* – the input data set must be normalised. The same procedures and considerations mentioned for normalising phase portraits apply here.
2 *Initialisation* – before training, the weight vectors are given initial, unique values (see Vesanto et al., 2000 for the PCA method).
3 *Find best-matching node* – each input vector is compared to the weight vectors of all nodes in the output. The node whose weight vector has the shortest Euclidean distance to the input at that step in the iteration is declared the best-matching node.

4 *Update weights* – the weights of the best-matching node are updated to be more similar to the input and thus, *better* represent the input – this is known as competitive learning.
5 *Update neighbourhood* – although there is only one best-matching node for any input vector, the nodes in a user-defined neighbourhood radius also get their weights updated, but in a decaying manner relative to their Euclidean distance to the best-matching node.
6 *Repeat iteration* – repeat steps 3–5 in two phases: rough training and fine-tuning (see Vesanto et al., 2000 for parameter settings) until errors level off.

The neighbourhood function (step 5) has the effect of clustering similar data onto similar regions of the output map, thus the term *self-organising*. Standard software packages, such as the SOM Toolbox for MATLAB (Vesanto et al., 2000), are available to easily run the SOM algorithm. The analyst should of course follow proper biomechanical data processing procedures to avoid erroneous results (see Chapter 9).

There are many ways to visualise the output layer beyond the simple linearly spaced grid shown in Figure 3.8 – which may give the impression of the nodes' weights being equally distributed. One popular method, known as the Unified distance matrix (U-matrix: Ultsch, 1993), is shown in Figure 3.9. The U-matrix in Figure 3.9 represents the lower body variables mentioned in the example earlier (left and right: ankle plantar-dorsiflexion, knee flexion-extension, hip flexion-extension, hip adduction-abduction and hip internal-external rotation) for over-ground walking. The nodes of the output layer in Figure 3.9 are shown as black dots in the centre of white hexagonal cells. Non-white hexagonal cells between the nodes are shaded to represent the similarity between neighbouring nodes. Dark cells represent relatively large Euclidean distances and lightly coloured cells represent relatively small Euclidean distances – or similar weight vectors – between neighbouring cells. The dark areas tend to represent borders and light areas, clusters. The black line drawn on Figure 3.9 is a trajectory connecting the best-matching nodes, consecutively for the entire duration of the movement. As can be seen in Figure 3.9, events and, therefore, phases of the movement can easily be identified. The trajectory winds around the dark central border, anticlockwise, although this is not necessarily always going to be true. Using the U-matrix visualisation, it is easy to see how variability between strides of the same or different individuals can be compared throughout the entire movement.

The SOM in Figure 3.9 was trained with the parameters shown in Table 3.1, all of which should be reported in publications and reports so that others may reproduce the output. The parameters shown are specific to the SOM Toolbox for MATLAB (Vesanto et al., 2000), for other software packages the parameter names and algorithms may differ slightly; therefore, the software and version used for the calculations should be identified. *Quantisation and topological errors* in Table 3.1 are measures of the quality of the map's fit to the input data. Quantisation error represents the average similarity between each input vector and its best-matching unit. Topological error measures whether the best-matching unit and the second-best-matching unit tend to be neighbours

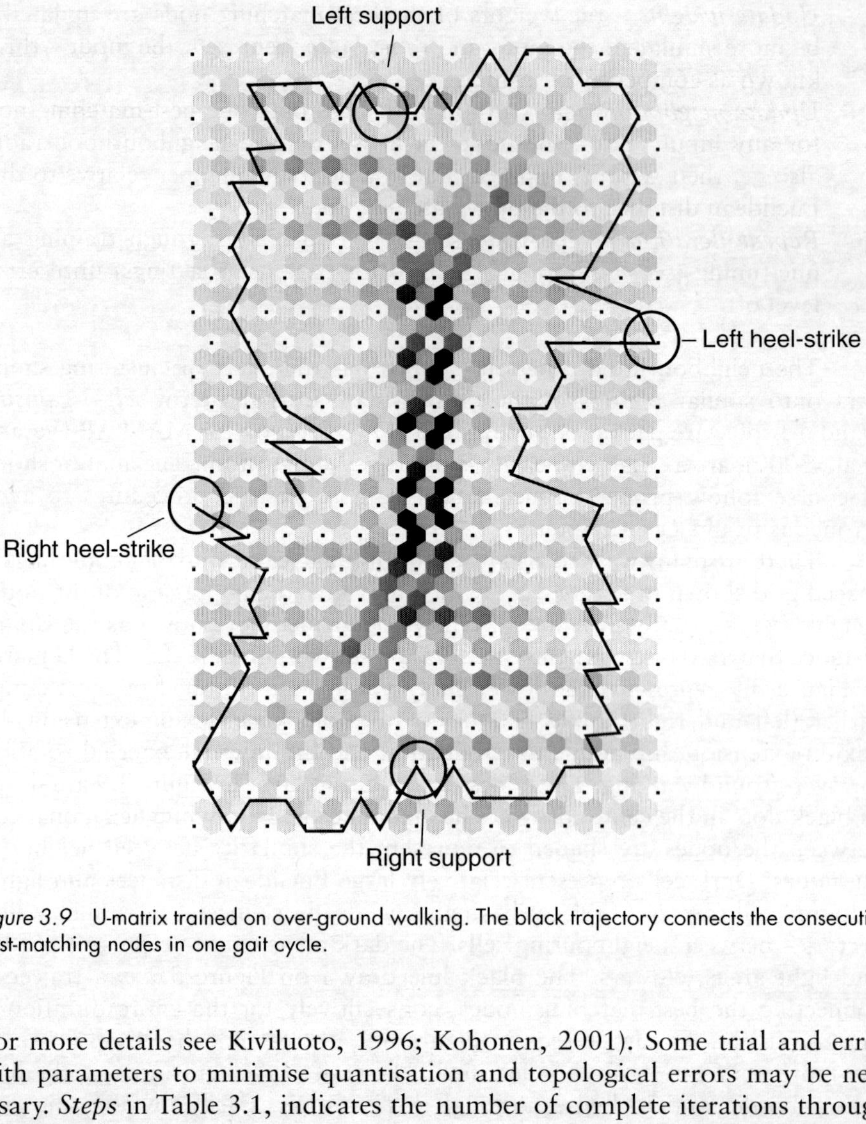

Figure 3.9 U-matrix trained on over-ground walking. The black trajectory connects the consecutive best-matching nodes in one gait cycle.

(for more details see Kiviluoto, 1996; Kohonen, 2001). Some trial and error with parameters to minimise quantisation and topological errors may be necessary. *Steps* in Table 3.1, indicates the number of complete iterations through the entire data set.

Looking at the movement this way allows the analyst a simple representation of the global coordination pattern – or the coordination of the lower body in our example. There are no assumptions to be satisfied, as is the case in statistics. However, a reasonably sized data set should be used for training to provide context to the visualisation. In other words, one trial could be passed to the SOM for training but the output map would be of little use.

Also note that more than one SOM can be used in a single analysis depending on the structure of the input data and the question being asked. In a study of basketball shooting by Lamb, Bartlett, and Robins (2010), one SOM was trained on the basketball shots of four players. Each player performed three different types of shot: the free throw shot, the three-point shot and the hook shot. Since there were two major sources of variability in the data set – shot type and individual differences – the resolution of the map was sufficient to see the

Table 3.1 SOM training parameters.

Parameter	*Value*
Software	SOM Toolbox for MATLAB, MATLAB r2012b (ver. 8), win64
Initialisation	Linear (PCA)
Lattice	Hexagonal
Neighbourhood function	Gaussian
Training algorithm	Batch
Map size	25 × 13
Rough training	
Steps	12
Radius	4 decreasing to 1
Fine-tuning	
Steps	200
Radius	1
Quantisation error	0.242
Topological error	0.041

differences in coordination associated with both sources of variability. On the other hand, in a study on golf chipping (Lamb et al., 2011) in which the type of shot was only slightly modified by small changes in target distance, the variability between individuals was much greater than the variability in coordination for the changing target distances. In this situation the single SOM clustering (not included in Lamb et al., 2011) was mainly of the individuals and the change in their coordination patterns at different target distances was masked. Since the research question in the golf chipping study was whether or not coordination changed under varying target distances, the authors chose to train four SOMs instead of one: one for each golfer in the study. Compared to the coarse differences between different types of basketball shot, the increased resolution was needed to see the subtle changes in coordination at different target distances. These decisions on whether to use a single SOM for the entire data set in a study, or analysis, or several SOMs must hinge on the questions being asked and the structure of the data.

REPORTING A COORDINATION STUDY

Information about the following should always be included when reporting a study of movement coordination.

- The exact definitions of the angles used and how they were measured.
- The accuracy and reliability of the angle measurements if these were assessed.

- The sampling frequency.
- Which normalisation procedures, if any, were used.

The following information specific to the analysis method used should also be reported.

- Angle–angle diagrams:
  - If the diagrams were quantified in any way and, if so, how.
- Cross-correlation functions:
  - Justification for each of the cross-correlations performed, in terms of the variables being meaningfully related.
  - If statistical significance is to be assigned to values of the cross-correlation function, whether relevant statistical assumptions were satisfied.
- Continuous relative phase:
  - How the phase angle was calculated.
  - How many trials were used in the analysis and whether the normalisation factors were gained from each trial or from a group of trials.
- Self-organising maps:
  - A list of all variables used in the analysis.
  - How many SOMs were used in the analysis and justification for the decision.
  - Training parameters, map dimensions, quality measures and software used.

REFERENCES

Alhoniemi, E., Himberg, J., Parviainen, J. and Vesanto, J. (2012) SOM Toolbox 2.1 for MATLAB [computer software]. https://github.com/ilarinieminen/ SOM-Toolbox.

Amblard, B., Assaiante, C., Lekhel, H. and Marchand, A. R. (1994) 'A statistical approach to sensorimotor strategies: conjugate cross-correlations', *Journal of Motor Behaviour*, 26(2): 103–112.

Bartlett, R. M. (2014) *Introduction to Sports Biomechanics: Analysing Human Movement Patterns*, 3rd edn. London: Routledge.

Bartlett, R. M., and Bussey, M. (2012) *Sports Biomechanics: Reducing Injury Risk and Improving Sports Performance*, 2nd edn. London: Routledge.

Bauer, H., and Schöllhorn, W. I. (1997) 'Self-organizing maps for the analysis of complex movement patterns', *Neural Processing Letters*, 5: 193–197.

Button, C., Davids, K. and Schöllhorn, W. (2006) 'Coordination profiling of movement systems', in K. Davids, S. Bennett. and K. Newell (eds), *Movement System Variability*, pp. 133–152. Champaign, IL: Human Kinetics.

Fuchs, A., Jirsa, V. K., Haken, H. and Kelso, J. A. S. (1996) 'Extending the HKB model of coordinated movement to oscillators with different eigenfrequencies', *Biological Cybernetics*, 74: 21–30.

Hamill, J., Haddad, M. and McDermot, W. J. (2000) 'Issues in quantifying variability from a dynamical systems perspective', *Journal of Applied Biomechanics*, 16: 407–418.

Hershler, C. and Milner, N. (1980) 'Angle-angle diagrams in the assessment of locomotion', *American Journal of Physical Medicine*, 59(3): 109–125.
Howell, D. C. (2010) *Statistical Methods for Psychology*, Independence, KY: Cengage Learning.
Kelso, J. A. S. (1995) *Dynamic Patterns: The Self-Organization of Brain and Behavior*, Cambridge, MA: MIT Press.
Kiviluoto, K. (1996) 'Topology preservation in self-organizing maps,' in *Proceedings of the International Conference on Neural Networks (ICNN'96)*, vol. 1, (Piscataway, New Jersey, USA), pp. 294-299, IEEE Neural Networks Council, June 1996.Kohonen, T. (2001) *Self-Organizing Maps*, 3rd edn. Berlin: Springer-Verlag.
Kurz, M. J. and Stergiou, N. (2002) 'Effects of normalisation and phase angle calculations on continuous relative phase', *Journal of Biomechanics*, 35: 269–374.
Kutilek, P. and Farkisova, B. (2011) 'Prediction of lower extremities' movement by angle-angle diagrams and neural networks', *Acta of Bioengineering and Biomechanics*, 13(2): 57–65.
Lamb, P. F., Bartlett, R. M., Lindinger, S. and Kennedy, G. (2014) 'Multi-dimensional coordination in cross-country skiing analyzed using self-organizing maps', *Human Movement Science*, 33: 54–69.
Lamb, P. F., Bartlett, R. M. and Robins, A. (2010) 'Self-organising maps: an objective method for complex human movement', *International Journal of Computer Science in Sport*, 9: 20–29.
Lamb, P. F., Bartlett, R. M. and Robins, A. (2011) 'Artificial neural networks for analyzing inter-limb coordination: the golf chip shot', *Human Movement Science*, 30: 1129–1143.
Lamb, P. F. and Stöckl, M. (2014) 'On the use of continuous relative phase: review of current approaches and outline for a new standard', *Clinical Biomechanics*, 29(5): 484–493.
Miller, R. H., Chang, R., Baird, J. L., van Emmerik, R. E. A. and Hamill, J. (2010) 'Variability in kinematic coupling assessed by vector coding and continuous relative phase', *Journal of Biomechanics*, 43: 2554–2460.
Morriss, C., Bartlett, R. M. and Fowler, N. (1997) 'Biomechanical analysis of the men's javelin throw at the 1995 World Championships in Athletics', *New Studies in Athletics*, 12: 31–41.
Peters, B., Haddad, J., Heiderscheit, B., van Emmerik, R. E. A. and Hamill, J. (2003) 'Limitations in the use and interpretation of continuous relative phase', *Journal of Biomechanics*, 36: 271–274.
Stergiou, N., Scholten, S. D., Jensen, J. L. and Blanke, D. (2001) 'Intralimb coordination following obstacle clearance during running: the effect of obstacle height', *Gait and Posture*, 13(3): 210–220.
Trezise, J., Bartlett, R. and Bussey, M. (2011) 'Coordination variability changes with fatigue in sprinters', *International Journal of Sports Science and Coaching*, 6(3): 357–364.
Ultsch, A. (1993) 'Self-organising networks for visualisation and classification', in O. Opitz, B. Lausen, and R. Klar (eds), *Information and Classification*, pp. 307–313. Berlin: Springer.
Vesanto, J., Himberg, J., Alhoniemi, E. and Parhankangas, J. (2000). SOM Toolbox for MATLAB 5 (No. A57). Helsinki, Finland: Helsinki University of Technology.
Wheat, J. S. and Glazier, P. S. (2006) 'Measuring coordination and variability in coordination', in K. Davids, S. Bennett, and K. Newell (eds), *Movement System Variability*, pp. 167–181. Champaign, IL: Human Kinetics.

CHAPTER 4

MOTION ANALYSIS USING VIDEO

Carl J. Payton and Christopher R. Hudson

INTRODUCTION

Over the last decade, considerable advances have been made in video technology. Most mobile phones and computer tablets now have a video recording function as standard with many models providing high definition quality images and some even offering high-speed recordings. Although these devices may provide images suitable for some forms of qualitative analysis, they as yet do not offer some key features present in video cameras that are required to perform a detailed quantitative analysis. Current entry-level video cameras give a picture quality that would have once been considered the domain of the professionals. Cameras providing full high definition quality are now available for £200 or less and the price of high speed models are many times lower than they were decades ago.

Video recordings of sport and exercise activities are usually made by biomechanists in order to undertake detailed analysis of an individual's movement patterns. Although on-line systems (Chapter 5) provide an attractive alternative to video as a method of capturing motion data, video motion analysis has a number of practical advantages over on-line motion analysis including:

- Low cost – video analysis systems are generally considerably cheaper than on-line systems.
- Minimal interference to the performer – video analysis can be conducted without the need for any disturbance to the performer, e.g. attachment of reflective markers.
- Flexibility – video analysis can be used in environments where some on-line systems would be unable to operate effectively, e.g. outdoors, underwater, in competition.
- Allows visual feedback to the performer – video cameras provide a permanent record of the movement that can be viewed immediately. On-line systems do not generally record the image of the performer.

Given the advantages just listed, video analysis will remain, for the foreseeable future, an important method of analysing movement in sport and exercise. Video analysis of an individual's technique may be qualitative or quantitative in nature. Qualitative analysis involves a detailed, systematic and structured observation of the performer's movement pattern (see Chapter 2). The purpose of this type of analysis is often to establish the quality of the movement being observed in order to provide some feedback to the performer. It may also be used as a means of identifying the key performance parameters that need to be quantified and monitored in future analyses.

The focus of this chapter is on quantitative analysis which involves taking measurements from video recordings to enable the quantification of selected movement variables in two dimensions (2D) or three dimensions (3D). The basic stages of any video-based quantitative analysis process are:

1 Video capture – recording video images of a calibration object and the performer.
2 Digitisation – obtaining 2D image coordinates of the calibration object and performer.
3 Smoothing – removing random error from the 2D image coordinates.
4 Reconstruction – converting 2D image coordinates to real-world coordinates (2D or 3D).
5 Calculation of biomechanical variables – obtaining relevant kinematic and kinetic variables from the real-world coordinates.

These stages require sophisticated hardware and software and it is important to follow the correct data capture and data processing procedures. Quantitative analysis can be time-consuming as it requires careful set-up and calibration of cameras and then often involves manual digitisation of multiple markers (typically 18 or more markers for a full body model) over a large number of video images. Depending on the type of model, the markers may be estimated joint centres of rotation (e.g. knee joint centre), segmental endpoints (e.g. end of foot), segment tracking markers or external objects (e.g. a sports implement). Two-dimensional image coordinates, resulting from the digitising process, are then smoothed and converted to real-world coordinates before being used to calculate linear and angular displacement-time histories. Additional kinematic information (velocities and accelerations) is obtained by computing the first and second time derivatives of these displacement data. The accuracy of these derivatives will be compromised unless appropriate data processing techniques are used (discussed in Chapter 9). Kinematic information extracted from video can be used to quantify key performance parameters (e.g. release angle during a throw). Such parameters can be compared between performers (e.g. novice vs. elite), within performers (e.g. fatigued vs. non-fatigued), or monitored over a period of time (e.g. to evaluate the effects of training over a season).

To understand the underlying causes of a sport or exercise technique, more detailed quantitative analyses are often performed. A common approach is that of inverse dynamics which involves computing kinetic information on the performer (e.g. net joint reaction forces and net moments) from kinematic information (see Chapter 11). Inverse dynamics equations require the linear and angular accelerations of the body segments being analysed and valid segment inertia data (mass and moment of inertia). The calculated joint moments and forces can have significant errors unless care is taken to minimise errors in the kinematic and inertia data.

The interpretation of data from an inverse dynamics analysis is not as straightforward as for a kinematic analysis. Inverse dynamics provides an insight into the net effect of all the muscles crossing a joint, but it does not allow direct computation of bone contact forces or the forces produced by individual muscles, or muscle groups. These variables can be inferred if internal joint parameters, such as joint centre location, muscles attachment sites, are known. Although the inverse dynamics approach has clear limitations (e.g. Winter, 1990), it can still provide the biomechanist with a much better understanding of the musculoskeletal forces and moments acting during a sport or exercise activity, than could be obtained from an analysis of the movement patterns alone.

EQUIPMENT CONSIDERATIONS

Selection of the appropriate equipment is important when undertaking a quantitative motion analysis study using video. The two main components of a video motion analysis system are

- Video camera and storage device – to capture and store video images of the movement;
- Computer system (hardware and software) – to digitise the video images, smooth and reconstruct coordinates and calculate biomechanical variables.

Video cameras

When selecting a video camera with the intention of undertaking a biomechanical analysis of a sport or exercise activity, the important features to consider are

- Picture quality
- Sampling frequency
- High-speed shutter/exposure time
- Manual iris and light sensitivity
- Synchronisation
- Recording medium

Picture quality

A video image is made up of a two-dimensional array of pixels. Video cameras capture an image using one of two methods: interlaced scan or progressive scan. When using the interlace technique a full video image or frame consists of two halves or fields. One field is made up of the odd-numbered horizontal lines of pixels, the other is made up of the even-numbered lines. The camera records one field first, followed by the second, and so on. With progressive scan, the camera records a complete frame. Many cameras now have the facility to capture images in either format. Thus, a camera recording 50p images will produce 50 full frame images per second. In contrast, on a 50i setting, a camera will record 50 fields per second and some form of image processing is performed to create a full video image during digitisation. Generally, given the choice, progressive scan should be selected as it will produce a higher quality image than the interlaced one.

The number and size of pixels making up a video image determine the resolution of the picture and this, to a large extent, dictates the picture quality. The vertical resolution refers to the number of horizontal lines in the video image; the horizontal resolution denotes the number of pixels per horizontal line. Many video cameras allow the user to specify the resolution of the image. When purchasing a video camera, it is important to check what modes the camera can record in and resolution(s) it can provide. Table 4.1 shows the main video recording formats that are currently available and their associated resolutions.

It should be noted that even within a given recording format, e.g. HD, the quality of the video image can vary considerably between cameras. One important factor is the size and quality of the camera's image sensor – the component that converts the light from the object into an electrical signal. The most common type of video image sensor is the complementary metal oxide semiconductor (CMOS). This sensor requires far less power than its predecessor, the charge-coupled device (CCD), and is now used in most standard and high-speed video cameras. In general, a large image sensor will give higher quality images than a small one.

Another important factor that will affect picture quality is the video file format: video data are usually stored in a compressed format to reduce the file size. Two important considerations here are the *container* (e.g. AVI, MP4, WMV, MOV) which holds the video data, and the *codec* (e.g. MJPEG, H.264) which is the software that encodes and decodes the video data. Note that a video file's extension, e.g. LongJump.AVI, designates the container, not the codec that has been used to encode the video in the file or the quality of the video data it contains. For motion analysis software to play a video file, the relevant codec must be installed on the computer being used for the analysis.

Table 4.1 Video formats and their associated resolutions.

	Horizontal resolution (pixels)	*Vertical resolution (lines)*
Standard Definition (SD)	720	576
High Definition (HD)	1280	720
Full High Definition (FHD)	1920	1080
Ultra High Definition (UHD)	3840	2160
2K	2048	1080
4K	4096	2160

The specification of the camera lens is another important factor in determining picture quality. Digital video cameras will have both an optical zoom range, e.g. 20× and a digital zoom range, e.g. 400×. It is important to note that once a camera is zoomed in beyond the range of its optical system, the picture quality will drastically reduce and will be unsuitable for quantitative analysis. Accessory telephoto lenses can be used to increase the optical zoom of a digital video camera and avoid this problem. They also allow the user to increase the camera-to-participant distance, whilst maintaining the desired image size. This will reduce the perspective error although it should be noted that the addition of a telephoto lens will reduce the amount of light reaching the camera's image sensor. It is important to check how well a telephoto lens performs at the limits of the optical zoom, as this is where image distortion will be most pronounced. Wide-angle lenses can be fitted to video cameras to increase the field of view for a given camera – participant distance. However, such lenses tend to produce considerable image distortion, which must be accounted for using specialised techniques when performing quantitative analyses.

Sampling frequency

The sampling frequency of a video camera, sometimes called frame rate, refers to the number of images captured per second. Until quite recently, most video cameras had a fixed sampling frequency of 25 frames per second. However, even entry level models now provide a range of sampling frequencies. For example, a low cost camera might offer the following progressive and interlaced sampling options: 24p, 25p, 30p, 50p, 50i, 60i, 60p. For some sport and exercise activities these frequencies will be too low and a high-speed video camera may be required. Although video cameras with sampling frequencies beyond 2000 Hz are commercially available, cameras with frequencies of 100–500 Hz are generally adequate for most sport and exercise biomechanics applications.

Most very high-speed video cameras (>1000 Hz) record their images to RAM, whereas less high-speed models record direct to a computer hard drive via a Firewire (IEEE1394), Ethernet or USB3 cable. One of the limitations of high-speed cameras that record to RAM is the limited recording time available. For example, the high-speed camera shown in Figure 4.1 has a storage capacity of 12 Gb. Recording in monochrome with a resolution of 1280 × 800 at 1630 Hz, it provides a recording duration of just 4.7 seconds. High-speed cameras that record to a computer via a cable are limited, not by recording time, but by the transmission speed of the cable and the computer hardware.

High-speed shutter/exposure time

For most biomechanics applications, a video camera equipped with a high-speed shutter is essential. The shutter is the component of a camera that controls the amount of time the camera's image sensor is exposed to light. This is referred to as the exposure time. Modern video cameras use electronic shuttering, which

Figure 4.1 Vision Research Phantom Miro 110 high-speed camera. Camera has a maximum sampling frequency of 1630 Hz at full resolution (1280 × 800). Higher sampling frequencies are achievable by reducing the resolution, e.g. 5090 Hz at 640 × 460 resolution.

involves activating or deactivating the image sensor for a specified time period, as each video image is sampled. When recording movement using a low shutter speed (long exposure time), the image sensor is exposed to the light passing through the camera lens for a relatively long period of time; this can result in a blurred or streaked image being recorded. The extent of the blurring would depend on the speed of the movement being analysed. Use of a shorter exposure time will reduce image blurring for a given movement but will also result in a darker image.

It is important that a video camera has a manual shutter speed function. This allows the user to select an exposure time that is appropriate for the activity being analysed, and the prevalent lighting conditions (see Data Collection Procedures section of this chapter). Typically, a video camera will offer exposure times ranging from 1/60–1/4000 s. It should be noted that not all video cameras offer a manual shutter function. Cameras that incorporate a Sports Mode function should be avoided because the exposure time associated with this is often inadequate for fast-moving activities.

Manual iris and light sensitivity

The iris is the element of the camera's lens system that regulates the amount of light falling on the image sensor by controlling thc aperture (the adjustable gap in the iris). If too much light is permitted to pass through the lens (large aperture), for too long, the result will be an overexposed image. If too little light passes through the lens (small aperture), the image will be underexposed.

Video cameras generally have automatic iris control that continually adjusts to ensure the image is correctly exposed. Some camera models have a manual override that allows the user to specify the iris setting. This is sometimes necessary when conducting biomechanical analyses. For example, when a short exposure time is needed in low light conditions, the aperture may have to be opened wider than it would be in automatic mode. The drawback of doing this is the increased noise level in the image, which results in a more 'grainy' picture. Additionally, the wider the aperture, the shallower will be the depth of field (distance from the plane of focus over which objects are still acceptably focussed). Video cameras each have a minimum light level that they require in order to produce an image. This level is expressed in lux. A camera with a minimum illumination value of 1 lux will perform better in low light conditions than one with 3 lux.

Synchronisation

For 3D video analysis (or 2D analysis where multiple cameras are needed), the activation of the camera shutters should ideally be perfectly synchronised so video images are scanned at precisely the same instants in time. One approach involves physically linking two or more camcorders with a gen-lock cable. Unfortunately, only a few expensive video camcorders have the facility to be gen-locked. An alternative approach is to use networked Ethernet cameras. With a trigger input cable, these cameras can be triggered remotely from a computer located 100 m or more away and the shutters perfectly synchronised. Software-based synchronisation of Ethernet cameras is possible without using the trigger input but small synchronisation errors may result.

If cameras cannot be synchronised during video capture, the 2D coordinates obtained from each camera view must be synchronised by interpolating the data and then shifting one data set by the time lag between the camera shutters. The time lag will be no more than half the reciprocal of the sampling frequency of the camera (e.g. <10 ms at 50 Hz). The simplest method of determining the time lag is to have a timing device in the field of view of all cameras. Where this is not possible, for example when filming at a competition, a method involving an analysis of the coordinates of all the digitised body markers has been proposed (Yeadon and King, 1999). Alternatively, certain commercial motion analysis software applications, for example SIMI Motion©, will automatically measure the time lag between camera shutters, if the video images from the cameras are simultaneously captured to a computer's hard drive, in real-time. The software will also interpolate and phase shift the 2D image coordinates to enable 3D coordinate reconstruction to be performed.

Recording medium

Recording video data on tape, including the popular MiniDV format, has become obsolete over the past few years. The two general options for recording video files are (1) locally within the camera (built-in memory or removable memory cards, e.g. SD cards) or (2) remotely on a computer. The advantages of recording locally are that the cameras tend to be self-contained units (most commonly camcorders) that are portable and easy to use in the field, as they do not require mains power or cabling to a computer. The advantages of remote recording are that multiple cameras can be controlled from a single location, greater control is available over the video recording format, video files are available immediately for analysis and camera synchronisation is often easier. Regardless of the method of recording used, the end result is a video file suitable for analysis on a computer.

Computer system (hardware and software)

When selecting and using a computer system with the intention of undertaking a video-based quantitative analysis of a sport or exercise activity, the important factors to consider are:

- Video format
- Capture hardware
- Display hardware
- Motion analysis software

Video format

It is important to ensure that the video file format on the computer (e.g. H.264 codec in an MP4 container) is compatible with the motion analysis software application. If the format provided by the camera is not compatible with the application, then it can be changed using widely available video conversion software (e.g. Prism, NCH Software). This conversion is a processor-intensive operation: the better the processor the less time the conversion will take.

Most motion analysis software applications will handle a variety of file formats and so the user has some decisions to make. The primary aim should be to minimise loss of video image quality during the initial capture and any subsequent conversion. One option is to capture video in an uncompressed format. Although this will give the highest possible visual quality, it will result in a large file size which may be impractical if recording long duration activities. If a file is to be compressed, a codec such as H.264, which is highly configurable, allows the user to select the combination of quality and file size that meets their needs.

Capture hardware

For cameras that use remote recording, the computer must capture a continuous stream of video data from the cameras and store it in a video file. The number of cameras and type of interface (e.g. USB3) will determine the capture hardware required. Each type of interface requires a specific port through which the computer receives the video data and each camera requires a separate port. If the required port or ports are not integral to the computer, then an expansion card with one or more ports can be attached via the computer's PCIe or ExpressCard slot.

Firewire (IEEE 1394) was a popular way of capturing video data from camcorders, but is now being replaced by HDMI. This requires a capture device, which sits between the camcorder and the computer (e.g. Blackmagic Designs DeckLink Studio 4K, Blackmagic Design Pty. Ltd.). It is important to check that the HDMI capture device supports the combination of video resolution and sampling frequency required. A USB camera can be connected to a USB3 port, which most modern computers have. Similarly, most computers typically have at least one Ethernet port, but it should be confirmed that it is a gigabit port (i.e. 10^9 bits per second), which most Ethernet cameras require. In addition, Ethernet cameras are often powered via the same cable that carries the video data; this is called Power over Ethernet (PoE) and often requires the inclusion of an inline PoE injector to power the camera.

Display hardware

The monitor is arguably the most critical hardware component when displaying video images for digitisation. Its display resolution should be at least the same as that of the video, for example, FHD video should be displayed on a monitor with at least 1920 × 1080 resolution.

To obtain 2D image coordinates it is necessary to advance the video image-by-image; playing it at normal speed is not required. For this application it is the computer's processor, rather than its video card, that decodes the video, so a better processor will provide smoother transition between images.

Motion analysis software

Some motion analysis software applications such as Dartfish (www.dartfish.com) are designed primarily for qualitative or semi-quantitative analyses, whereas applications such as Quintic Biomechanics (Quintic Consultancy Ltd.) and SIMI Motion© (Simi Reality Motion Systems GmbH) provide support for all aspects of the video-based quantitative analysis process (video capture, digitisation, smoothing, reconstruction and calculation of biomechanical variables).

An important consideration when selecting analysis software is the smallest measurement increment that can be detected on the video image during the digitising process, that is, the digitiser resolution. This affects the level of precision to which the coordinates can be measured. Current video-based digitising systems offer high measurement resolutions through a combination of zoom and a sub-pixel cursor. It should be noted that unless the resolution of the captured video is high, the image will become very 'pixilated' at high magnifications.

Most video analysis software now provides automatic tracking of body markers as standard. This can save considerable digitisation time but requires highly contrasting markers that remain clearly visible in all images. Applications are also now commercially available that provide full-body markerless tracking based on silhouettes of the body (see for example SIMI Shape 3D, Simi Reality Motion Systems GmbH).

Good motion analysis software should provide a wide range of data smoothing options (see Chapter 9), a choice of 2D and 3D reconstruction methods (see next section), and the facility to calculate all the standard kinematic variables. Some packages provide further, more advanced options such as inverse dynamics and segmental energy transfer analysis. It is important that the software permits data to be imported and exported in various formats compatible with other software, for example as a C3D file which is widely supported by other software tools such as Visual3D (C-motion, Inc.).

DATA COLLECTION PROCEDURES

When conducting a quantitative video analysis certain procedures must be followed carefully, at both the video recording and digitising stages, to minimise the systematic and random errors in the digitised coordinates. Even when undertaking a qualitative video analysis, many of the video recording procedures are still pertinent as they will help to obtain a high quality video record of the performance.

Quantitative video analysis may be two dimensional (2D) or three dimensional (3D). The former approach is much simpler, but it assumes that the movement being analysed is confined to a single, pre-defined plane – the plane of motion. Any measurements taken of movements outside this plane will be subject to perspective error, thus reducing their accuracy. Even activities that appear to be two dimensional, such as a walking gait, are likely to involve movements in more than one plane; a 2D analysis would not enable these to be quantified accurately. Three-dimensional analysis enables the true spatial movements of the performer to be quantified. This approach eliminates perspective error but the recording and analysis procedures are more complicated, and the equipment requirements are also greater.

Two-dimensional video recording

The following guidelines are designed to minimise the systematic and random errors present in 2D coordinates, resulting from the video recording stage. This will increase the accuracy of any parameters subsequently obtained from these coordinates. The guidelines are based on those previously reported in Payton and Bartlett (2008) and in many earlier texts.

Equipment set-up

Mount the camera on a stable tripod and avoid panning

The standard approach in a 2D analysis is for the camera to remain stationary as the performer moves through the field of view. This enables the movement of the performer to be determined easily relative to an external frame of reference. Two-dimensional filming techniques involving panning or tracking cameras have been used when the performance occurs over a long path (e.g. Gervais et al., 1989; Chow, 1993). As these methods involve the camera moving (rotating or translating) relative to the external frame of reference, mathematical corrections have to be made for this movement if accurate 2D coordinates are to be obtained.

Maximise the camera-to-subject distance
The camera must be positioned as far as is practically possible from the performer. This will reduce the perspective error that results from movement outside the plane of performance (see Figure 4.2). A telephoto zoom lens will enable the camera-to-subject distance to be increased whilst maintaining the desired image size. A digital video camera should not be positioned beyond the limit of its optical zoom system for the reasons outlined earlier.

Maximise the image size
To increase digitising accuracy, the image of the performer must be as large as possible. Image size is inversely proportional to the field of view of the camera so the camera should only be zoomed out sufficiently to encompass the performance path, plus a small margin for error. For events that occur over long performance paths, e.g. triple jump, a single stationary camera would not provide an image size suitable for quantitative analysis. Here, the use of multiple synchronised cameras, or a panning/tracking camera method, would be required.

Focus the camera manually
Most video cameras have an automatic focus system that can be manually overridden. In most situations, the camera should be set in manual focus mode. For a well-focused image, zoom in fully on an object in the plane of motion, manually focus and zoom out to the required field of view.

Calibrate the camera
A calibrated camera is needed for reconstruction, i.e. the conversion of 2D image coordinates to real-world coordinates. Calibrating a camera usually

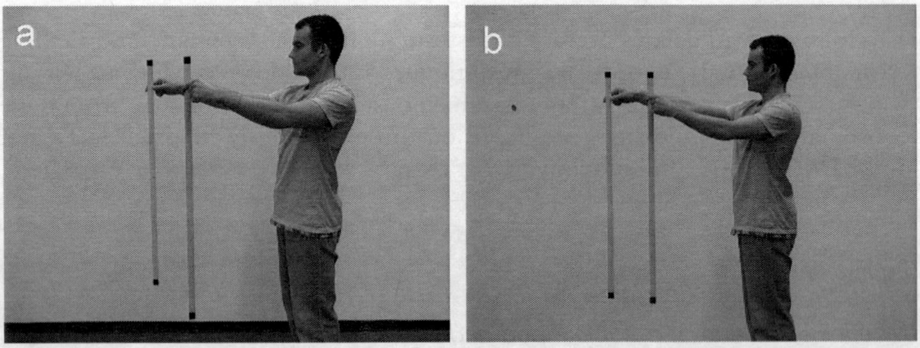

Figure 4.2 Apparent discrepancy in the lengths of two identical rods when recorded using a camera-to-subject distance of 3 m (image a) and 20 m (image b). Note that the rods are being held shoulder width apart.

requires video images of a calibration object or control points, which must be filmed using the same camera set-up as that used to film the performers. Two commonly used two-dimensional calibration methods are scaling object and 2D-DLT. A third, called planar, is useful if a wide field of view is required.

Method 1 – Scaling object: this approach requires the camera's image sensor to be parallel to the pre-defined plane of motion, i.e. perpendicular to the camera optical axis. As no human movement is truly planar, it is essential to establish which aspect of the activity is of primary interest and in which plane this occurs. The camera can then be positioned accordingly. Marking a straight line from the camera lens to the geometric centre of the field of view can represent the direction of the optical axis. Various methods can be used to align the optical axis orthogonal to the plane of motion. A common approach is to use right angle triangles (triangles whose sides are in the ratio 3:4:5). Failure to ensure that the optical axis is orthogonal to the plane of motion, even by a few degrees, can have a detrimental effect on the accuracy of the analysis (Brewin and Kerwin, 2003). Even with a correctly aligned camera, movement will inevitably occur outside the plane of motion. The effect on measured angles is illustrated in Figure 4.3.

To enable a true vertical (and horizontal) frame of reference to be established, a clear vertical reference, such as a plumb line, must be recorded. The video motion analysis system will usually allow correction for a non-vertically aligned camera, using the coordinates of the vertical reference.

An object whose dimensions are accurately known must be placed in the plane of motion and filmed. The object should be placed both horizontally and vertically, because the computer may display the image with an aspect ratio (i.e. the ratio of the width to the height) that distorts it in one dimension. To minimise the error in the scaling process the dimensions of the scaling objects should be such that they occupy a good proportion of the width and height of the field of view. For a given digitising error, the scaling error will be inversely proportional to the length of the scaling object. For field widths greater than 2–3 m, scaling is usually done using the known distance between two or more control points, positioned in the plane of motion.

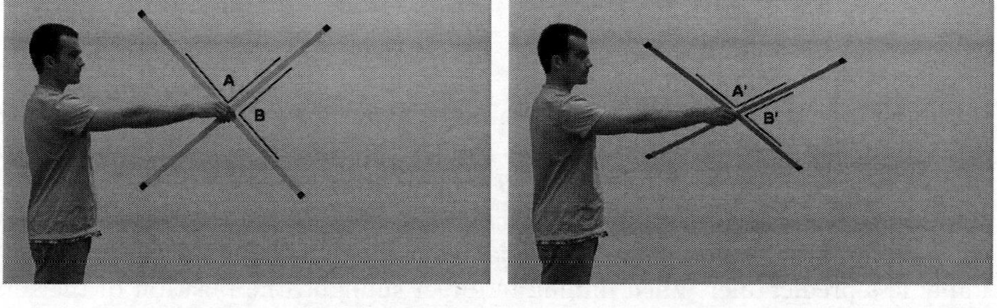

Figure 4.3 Distortion of angles when movement occurs outside the plane of motion. The true value of angles A and B is 90° (image a). In image b, angle A appears to be greater than 90° (A′) and angle B appears to be less than 90° (B′), as the frame is no longer in the plane of motion.

Method 2 – 2D-DLT: in some circumstances it is not possible to align the camera's optical axis to be perpendicular to the plane of motion, for example when filming in a competition. Here, digitisation of a grid of control points, placed in the plane of motion, can be used to correct for the camera misalignment. This method is called 2D-DLT and has been shown to provide significantly more accurate reconstruction than the more commonly used scaling object approach, particularly when the optical axis of the camera is tilted more than a few degrees relative to the plane of motion (Brewin and Kerwin, 2003).

Method 3 – Planar: for a wide field of view, the planar calibration approach (Zhang, 2000) can be used. A wide field of view can result in considerable lens distortion, which must be accounted for to give accurate reconstructions. The planar approach does this by filming a calibration object as it is moved through the camera's view. This object is typically a flat plate with a black and white checkerboard pattern on it. The resulting video images are then processed to ascertain the effect that the lens and any imperfections in the camera's construction had on the images. This method has been shown to have higher reconstruction accuracy than 2D-DLT for wide fields of view (Dunn et al., 2012). Software implementations of this method are freely available (e.g. Check2D, 2016 and Bouguet, 2016).

Select an appropriate exposure time and aperture

In activities such as running, jumping, throwing and kicking, it is the most distal body segments, the hands and feet, which move the quickest. An exposure time should be selected that is sufficient to provide a non-blurred image of the fastest moving body segments (or sports implements). The choice of exposure time depends on the activity being recorded. For slow movements, such as a grande plié in ballet or walking, exposure times of 1/150–1/250 s should be adequate; for moderately fast activities, such as running or a swimming start, exposure times of 1/350–1/750 s are more appropriate; for fast activities such as a golf swing or a tennis serve, an exposure time of 1/1000 s or above may be needed. The effect of using different exposure times on the recording of a tennis ball impacting a racket is shown in Figure 4.4.

An increase in exposure time will enhance the brightness of the recorded image, but also the potential for blurring, for given lighting conditions and camera aperture setting. To obtain the best possible images at the required exposure time, sufficient lighting must be provided such that the camera iris aperture does not have to be opened excessively.

Ensure correct lighting

If filming indoors, floodlights are often needed to achieve the required lighting level. Bartlett (1997) suggests that one floodlight positioned perpendicular to the plane of performance, and one to each side at around 30° to the plane, should provide adequate illumination. Filming outdoors in natural daylight is often preferable to filming under artificial lights, but natural light levels are inevitably less predictable. When filming in direct sunlight, the position of the sun will restrict where the camera can be located. The background must provide a good contrast with the performer and be as plain and uncluttered as possible. When filming indoors with floodlighting, a dark, non-reflective background is

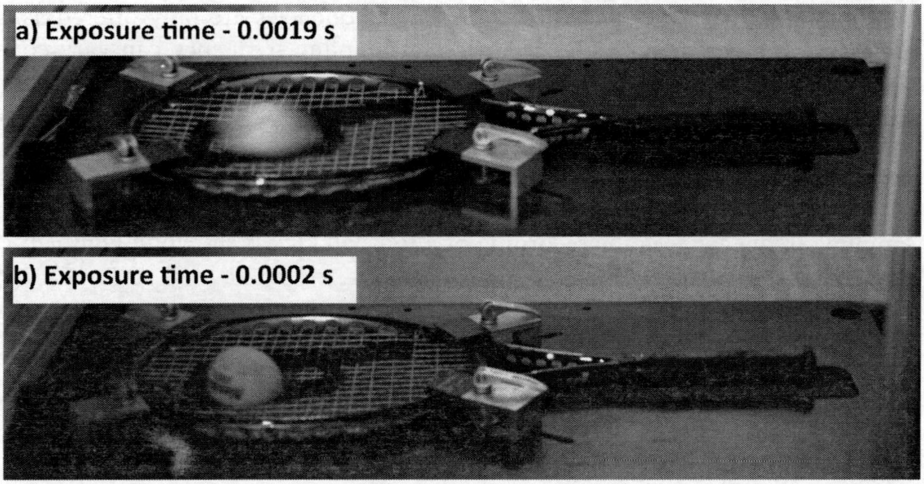

Figure 4.4 Effect of exposure time on the image quality of a tennis ball impacting a tennis racket at 90 mph (40 m s^{-1}). Images were recorded using a Vision Research Phantom Miro 110 high speed video camera (Figure 4.1) sampling at 500 Hz. The exposure times in images a and b were 0.0019 s and 0.0002 s, respectively.

preferred. Video cameras often have a manually adjustable setting for different light sources (e.g. daylight, fluorescent lamps, sodium or mercury lamps) and white balance, which can be used to enhance the colour rendition.

Select an appropriate sampling frequency

Standard video camcorders generally have a maximum frequency of 50–60 Hz whereas high-speed video cameras can sample at 1000 Hz and above. Sampling frequency selection should be based on the frequency content of the movement being analysed and the dependent variables being studied. Sampling theorum (see Chapter 9 for more detail) states that the sampling frequency must be at least double that of the highest frequency present in the activity itself. In reality, the sampling frequency should be much higher than this (Challis et al., 1997, suggest 8–10 times higher).

A sufficiently high sampling frequency will ensure that the instances of maximum and minimum displacement (linear and angular) of a joint or limb, and of other key events in a performance (e.g. heel-strike in running, ball impact in a golf swing) are recorded. An increase in the frequency will also serve to improve the accuracy of temporal measurements, for example, the phase durations of a movement. This is particularly important where the phases are of short duration, for example, the hitting phase of a tennis serve. Some suggested sampling frequencies for various activities are:

- 25–50 Hz – walking, swimming, stair climbing
- 50–100 Hz – running, shot put, high jump
- 100–200 Hz – sprinting, javelin throwing, football kick
- 200–500 Hz – tennis serve, golf swing, parry in fencing

It should be noted that these sampling frequencies are only offered as a guide. For a given activity, the appropriate sampling frequency can vary considerably depending on the measurements required. For example, a quantitative analysis of the interaction between the player's foot and the ball during a football kick would require a sampling frequency above 1000 Hz, whereas a frequency of 25 Hz would be more than adequate for determining the length of the final stride during the approach to the ball. The effect of using different sampling frequencies on the recording of a football kick is shown in Figure 4.5.

Participant preparation and recording trials

The health and safety of the participant is paramount. Informed consent should always be obtained from the participant and completion of a health questionnaire is often required. Sufficient time must be allocated for a warm-up and for the participant to become fully familiar with the testing environment and testing conditions.

The participant's clothing should allow the relevant limbs and body landmarks to be seen clearly. Careful placement of small markers on the skin can help the analyst to locate body landmarks during digitising, but marker position must be considered carefully. Movement of soft tissue means that surface markers can only ever provide a guide to the structures of the underlying skeleton. Markers are often used to help identify the location of a joint's instantaneous centre of rotation. Whilst a single marker can adequately represent the axis of a simple hinge joint, more complex joints will require more complex marker systems (see Chapter 5 for more detail on marker systems).

The number of trials recorded will depend on the purpose of the analysis and the skill level of the participants. As the movement patterns of skilled performers are likely to be significantly more consistent than those of novice performers (Williams and Ericsson, 2005), fewer trials may be necessary to capture a typical performance. During the filming it is often useful to record a board in the field of view, showing information such as the date, performer, trial number and condition, and camera settings.

Three-dimensional video recording

Many of the procedures described in the previous section for 2D video analysis will also apply when using a 3D approach. This section will discuss the main additional issues to be considered at the recording stage of a three-dimensional analysis.

Equipment set-up

Calibrate the cameras

The essential requirement is to have two or more cameras simultaneously recording the performance, each from a different perspective. A number of approaches

Figure 4.5 The effect of camera frame rate on the recording of a football kick. At 50 Hz (top row) the foot is only seen in contact with the ball for one image; at 250 Hz (middle row) the foot remains in contact for four images; at 1000 Hz (bottom row) the foot is in contact for 16 images (not all images shown).

are available for reconstructing real-world 3D coordinates from the 2D image coordinates. The two most commonly used methods are (1) the Direct Linear Transformation (DLT) (Abdel-Aziz and Karara, 1971), and (2) Wand Calibration (Borghese et al., 2001).

The DLT method determines a linear relationship between the 2D image coordinates of, for example, a body landmark, and its 3D real-world coordinates. To establish this relationship, an object space or performance volume must be defined using a set of control points whose real-world 3D coordinates are known. This is usually achieved using a rigid calibration frame of known dimensions incorporating a set of visible markers (see Figures 4.6). Alternatively, a series of discrete calibration poles can be used, provided their real world coordinates have been accurately established using, for example, surveying techniques. There is some debate as to whether filling the performance volume with control points improves DLT reconstruction accuracy compared to positioning the control points around the outside of the performance volume (see Challis and Kerwin, 1992; Chen et al., 1994).

A minimum of six non-coplanar control points is required for the reconstruction of 3D real-world coordinates, but 15–20 control points or more is recommended and many more may be needed for large performance volumes. The control point coordinates must be known relative to three orthogonal, intersecting

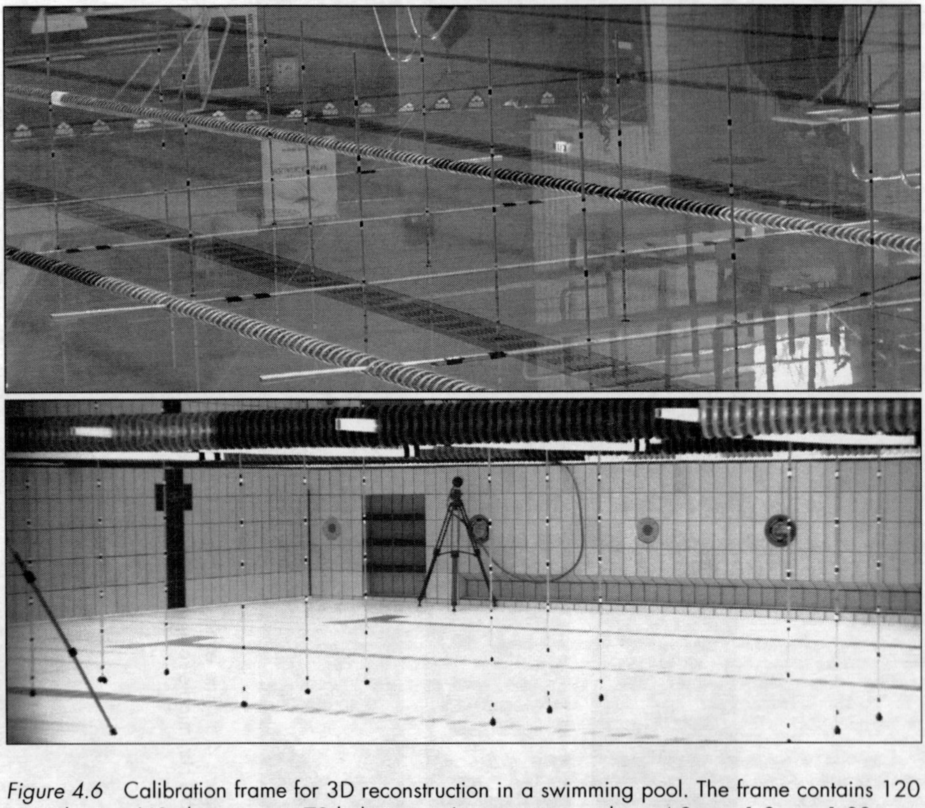

Figure 4.6 Calibration frame for 3D reconstruction in a swimming pool. The frame contains 120 control points (48 above water; 72 below water) occupying a volume 4.2 m × 1.0 m × 1.80 m.

axes, which define a global coordinate system or inertial reference system. This reference system is fixed in space and all 3D real-world coordinates are expressed relative to this. Video images of the control points are recorded by each camera which are then digitised to produce a set of 2D image coordinates for each control point from each camera view. These are used to compute eleven DLT parameters which relate to the orientation and position of each of the cameras (a detailed theoretical background to the DLT algorithm can be found elsewhere, e.g. Abdel-Aziz and Karara, 1971; Miller, Shapiro and McLaughlin, 1980).

For the minimum of six control points, 12 equations are produced for each camera view. As there are more equations than unknowns, the DLT parameters are obtained by solving the equations using a least-squares technique (Miller, Shapiro and McLaughlin, 1980). With the DLT parameters obtained, the same equations can then be used to obtain the 3D real-world coordinates of any marker in the object space, provided the 2D image coordinates of the marker are known from at least two of the cameras.

The calibration frame approach conveniently provides all the required calibration data from a single video image and is a highly reliable method of calibrating the cameras (Chen, Armstrong and Raftopoulos, 1994). However, the wand method provides an attractive alternative. First, this method involves defining the global coordinate system and initial camera parameter values using either a rigid L-frame or calibration wand (Figure 5.1), constructed with a number of control points of known location (see Chapter 5 for more detail). Then, the user walks through the required performance volume sweeping the calibration wand through as many locations and orientations as possible in, typically, 60 s (Pribanić et al., 2009).

The known distance(s) between the wand's control points and the 2D image coordinates of the wand, obtained from each camera, are then used to establish the camera parameters required for 3D reconstruction. The obvious advantages of the wand method over a calibration frame-based approach are the flexibility it offers in defining the performance volume size, the reduced set-up time, and the easier manufacture, storage and transportation of the wands. Regardless of the calibration method used, the calibrated volume must be large enough to encompass the movement being analysed as accuracy of the coordinates drops when movement occurs outside the calibrated volume (Hinrichs and McLean, 1995).

When setting up the equipment for three-dimensional analysis the biomechanist should follow the steps in the following sub-sections.

Mount the cameras on stable tripods and avoid panning

The standard approach in a 3D analysis is for the cameras to remain stationary as the performer moves through their field of view. Three-dimensional filming techniques involving panning cameras (e.g. Yu et al., 1993; Yanai et al., 1996); panning and tilting cameras (e.g. Yeadon, 1989); and panning, tilting and zooming cameras (e.g. Ai and Bi, 2013) have been used when the performance occurs over a long path. As these methods involve the cameras moving relative to the global coordinate system, a number of fixed reference markers have to be digitised in each video image to correct for the changing orientation of the cameras. Pan-Tilt-Zoom modules are now available within commercial motion analysis software (e.g. SIMI Motion©).

Position the cameras for optimum viewing of body markers
Body markers to be digitised must remain in view of at least two cameras for the duration of the activity. It is advisable to use as many cameras as is practically possible to help achieve this. Try to avoid having cameras facing each other or facing in a similar direction (optical axes parallel) as these camera arrangements will not be suitable for obtaining 3D coordinates. Typically, angles of 60–120° between camera optical axes should be the goal, bearing in mind that this can be achieved by positioning cameras at different heights.

Poorly positioned cameras may force the analyst into guessing marker positions at certain stages of the movement, inevitably reducing the accuracy of the coordinate data. Many video analysis programmes offer an interpolation function that can predict the position of obscured markers. This option should only be used when the marker is concealed for no more than four or five images and is not reaching a turning point (maximum or minimum) during that period.

Ensure control points are visible to and recorded by all cameras
When using the DLT method, all control points used to compute the DLT parameters must be clearly visible to each camera during calibration. When using a calibration frame, care should be taken to avoid the poles at the rear of the frame being obscured by those in the foreground. A good contrast between the control points and the background is also essential. It is strongly recommended that the control points are recorded at the start and the end of the data collection session. This will allow recalibration if one of the cameras is moved slightly during the session.

Align the performance with the axes of the global coordinate system
The International Society of Biomechanics (ISB) recommends that, where there is an obvious direction of progression, for example in gait, the X-axis of the global coordinate system be nominally aligned with this. They propose the use of a right-handed coordinate system, with the Y-axis being directed vertically and the Z-axis laterally.

Make provision for shutter synchronisation and event synchronisation
Ideally, all the cameras should have their shutters synchronised. Where this is not possible, the time lag between each of the camera shutters must be determined so the 2D image coordinates obtained from each camera view can be synchronised (see Video Cameras section in this chapter).

During filming it is useful to activate an event marker, such as an LED or strobe light, which is visible to all cameras. Such a device can be used to confirm that the first video image digitised from a given camera view corresponds temporally to the first image digitised from all other camera views. Failure to fulfil this requirement will result in erroneous 3D real-world coordinates.

Video digitising

The process of obtaining 2D image coordinates from a video record or 'digitising' may be achieved automatically or manually. Where skin-mounted markers are being used only as a guide to the location of an anatomical landmark, such as the gleno-humeral joint centre, manual digitisation may be required. The process of visually identifying and then marking the points of interest, frame-by-frame, will inevitably introduce some systematic and random errors to the coordinate data (see Chapter 9 for more detail). With attention to detail, these errors can be kept to an acceptable level. The following points should be considered when manually digitising a video sequence:

- The same operator should digitise all trials in the study to ensure consistency (reliability) between trials.
- Only ever use skin-mounted markers as a guide. Consider carefully the anatomical landmark being sought. A sound knowledge of the underlying musculoskeletal system is essential here.
- Great care should be taken when digitising the scaling object or control points. Any measurement error here will introduce a systematic error in the coordinate data, and in all variables derived from these.
- On completion of a 3D calibration, check that the 3D reconstruction errors fall within acceptable limits. These errors will depend mainly on the volume of the object-space being calibrated, the quality of the video image and the resolution of the digitiser. As a guide, Sanders et al. (2006) reported mean RMS reconstruction errors of 3.9 mm, 3.8 mm and 4.8 mm for the x, y and z coordinates, respectively, of 30 points distributed throughout a calibration frame measuring 1.0 m × 1.5 m × 4.5 m.
- A representative sequence should be digitised several times by the investigator to establish the intra-operator reliability. Inter-operator reliability (objectivity) should also be determined by having one or more other experienced individuals digitise the same sequence (see Atkinson and Nevill, 1998, for more information on assessing measurement reliability).

PROCESSING, ANALYSING AND PRESENTING VIDEO-DERIVED DATA

The digitising process will introduce high frequency errors (noise) to the 2D image coordinates. Essentially, what is required next is to (1) smooth the coordinates to remove the random errors and transform them to a form suitable for computing kinematic variables, and (2) calculate and display the kinematic variables in a format that allows the user to extract the information required to complete the analysis.

Smoothing and transforming coordinates

Various smoothing methods can be used to remove the high frequency noise; these fall into three general categories: digital filters, spline fitting and Fourier series truncation (Bartlett, 2014). Failure to smooth coordinates sufficiently will lead to high levels of noise in any derived kinematic variables, particularly acceleration. Over-smoothing of the coordinates will result in some of the original signal being lost. Selecting the correct smoothing factor for a given data set is therefore critically important. Chapter 9 provides a detailed discussion of smoothing methods and presents some practical guidelines for their use. The transformation of image coordinates to real world coordinates is necessary before any analysis can be undertaken. Procedures for achieving this were discussed earlier in this chapter.

Calculating kinematic variables

The sport and exercise biomechanist is often interested both in the movement patterns of individual body segments, for example in throwing and kicking, and in the overall motion of the performer's centre of mass, for example in gait analysis. Computation of the mass centre location requires a linked-segment model to be defined, and the mass, and mass centre locations, of individual body segments to be determined. Three general methods are used to obtain body segment parameters: regression equations based on measurements taken from cadavers, geometric modelling of the body segments and the use of imaging devices. More detail on these methods can be found in Durkin (2008). The biomechanist should seek to use segment inertia data that closely match the physical characteristics of the participants being analysed.

The linear displacement of a body landmark (or mass centre) in one dimension (e.g. x direction) is defined as the change in the relevant scaled coordinate of that landmark (Δx) during a specified time period. Resultant linear displacements in two or three dimensions are easily calculated using Pythagoras' theorem. Two-dimensional (planar) angles are obtained from 2D coordinates using simple trigonometry and may be relative (e.g. joint angles formed by two adjacent segments) or absolute (e.g. the angle of a segment relative to the vertical). Planar angles are quite straightforward to interpret once the angular conventions adopted by the analysis system have been established. The calculation of relative (joint) angles from 3D coordinates is more complex, as is their interpretation. The most common methods for calculating 3D joint angles in biomechanics are the Euler, Cardan and Joint Coordinate System (JCS) methods. Some video analysis software packages do not do these calculations so it may be necessary to export 3D coordinates into another analysis tool, such as Visual3D, to generate the angles. A detailed discussion of 3D angle calculation is provided in Chapter 9 and elsewhere (e.g. Andrews, 1995).

Linear and angular velocities and accelerations are defined as the first and second time derivatives of the displacement (linear or angular), respectively. These can be computed either numerically (e.g. finite difference method) or analytically (if the data have been smoothed with mathematical functions such as splines). As with displacement, the orthogonal components of velocity and acceleration can be analysed separately, or their resultants found.

Analysing and presenting video-derived data

In any biomechanical analysis the selection of dependent variables will be dictated by the aim of the study. It is important to identify the biomechanical variables of interest before commencing data collection, as this will influence the methodology used (e.g. 2D vs. 3D; normal vs. high-speed video). When analysing a sport or exercise activity the use of deterministic models (Chow and Knudson, 2011) can help to identify the important movement parameters, as can reference to the appropriate research literature.

There are a number of ways of presenting kinematic data from a video analysis and it is for the individual to decide on the most appropriate presentation format. This will be dictated mainly by the intended destination of the information (e.g. research journal, athlete feedback report). The most common methods of presenting kinematic data are as discrete measures (e.g. peak joint angles) and as time-series plots (e.g. hip velocity vs. time). Where the focus of the analysis is on movement coordination, the use of angle–angle plots and angle–angular velocity (phase) plots is becoming increasingly popular as discussed in Chapter 3.

REPORTING A VIDEO MOTION-ANALYSIS STUDY

The biomechanist should consider including some or all of the following information when reporting a video-based study.

Participants

- Participant details (age, height, body mass, trained status, etc.);
- Method of obtaining informed consent (verbal or written);
- Nature of the warm-up and familiarisation;
- Type of clothing worn, number, type and precise location of body markers.

Video capture

- Camera and lens type (manufacturer and model) and the recording medium, format and resolution (e.g. Full HD 1080p onto an SD Card);
- Camera settings (sampling frequency, exposure time, iris (f-stop) setting);
- Position of camera(s) relative to the movement being recorded and the field width or performance volume covered by each camera (a diagram is useful here);
- Method used to synchronise the cameras with each other (and with other data acquisition systems if used);

- Details of lighting (e.g. position of floodlights);
- Details of 2D or 3D calibration object (dimensions of object, number and location of control points).

Video digitising

- Capture and display hardware and software (manufacturer and model/version);
- Resolution of the digitising system;
- Digitising rate (this may be lower than the camera's sampling frequency);
- Model being used (e.g. 15 point segmental).

Processing, analysing and reporting

- Algorithm used to obtain the 3D coordinates;
- Method used to smooth the coordinates including a justification for the method and level of smoothing used;
- Method used to obtain the derivative data (e.g. numerical, analytical);
- Source of segment inertia data used to calculate e.g. the whole body mass centre or a limb moment of inertia;
- Definitions of all dependent variables used, including their SI units;
- Estimation of the measurement error in the calculated parameters;
- Level of inter- and intra-observer reliability of the calculated parameters.

REFERENCES

Abdel-Aziz, Y. I. and Karara, H. M. (1971) 'Direct linear transformation from comparator co-ordinates into object space co-ordinates in close range photogrammetry', in *American Society of Photogrammetry Symposium on Close Range Photogrammetry*, pp. 1–18. Falls Church, VA: American Society of Photogrammetry.

Ai, K. and Bi, Z. (2013) '3D image analysis with pan/tilt/zoom and its assessment of accuracy', in 31st International Conference on Biomechanics in Sports Conference Proceedings Archive. [online]. https://ojs.ub.uni-konstanz.de/cpa/article/view/5568 (accessed 9 August 2016).

Andrews, J. G. (1995) 'Euler's and Lagrange's equations for linked rigid-body models of three-dimensional human motion', in P. Allard, I. A. F. Stokes and J. P. Blanchi (eds), *Three-Dimensional Analysis of Human Movement*, pp. 145–175. Champaign, IL: Human Kinetics.

Atkinson, G. and Nevill, A. M. (1998) 'Statistical methods for assessing measurement error (reliability) in variables relevant to sports medicine', *Sports Medicine*, 26(4): 217–238.

Bartlett, R. M. (ed.) (1997) *Biomechanical Analysis of Movement in Sport and Exercise*, Leeds: British Association of Sport and Exercise Sciences.

Bartlett, R. M. (2014) *Introduction to Sports Biomechanics: Analysing Human Movement Patterns*, 3rd edn. London: Routledge.

Borghese, N. A., Cerveri, P. and Rigiroli, P. (2001) 'A fast method for calibrating video-based motion analysers using only a rigid bar', *Medical and Biological Engineering and Computing*, 39(1): 76–81.

Bouguet (2016) Camera calibration toolbox for MATLAB. [online]. www.vision.caltech.edu/bouguetj/calib_doc/.(accessed 14 June 2016).

Brewin, M. A. and Kerwin, D. G. (2003) 'Accuracy of scaling and DLT reconstruction techniques for planar motion analyses', *Journal of Applied Biomechanics*, 19: 79–88.

Challis, J., Bartlett, R. M. and Yeadon, M. (1997) 'Image-based motion analysis', in R. M. Bartlett (ed.) *Biomechanical Analysis of Movement in Sport and Exercise*, pp. 7–30. Leeds: British Association of Sport and Exercise Sciences.

Challis, J. H. and Kerwin, D. G. (1992) 'Accuracy assessment and control point configuration when using the DLT for photogrammetry', *Journal of Biomechanics*, 25(9): 1053–1058.

CHECK2D (2016) Check2D – camera calibration tool for 2D kinematic analysis. [online]. www.check2d.co.uk/. (accessed 14 June 2016).

Chen, L., Armstrong, C. W. and Raftopoulos, D. D. (1994) 'An investigation on the accuracy of three-dimensional space reconstruction using the direct linear transformation technique', *Journal of Biomechanics*, 27: 493–500.

Chow, J. W. (1993) 'A panning videographic technique to obtain selected kinematic characteristics of the strides in sprint hurdling', *Journal of Applied Biomechanics*, 9: 149–159.

Chow, J. W. and Knudson, D. V. (2011) 'Use of deterministic models in sports and exercise biomechanics research', *Sports Biomechanics*, 10: 219–233.

Dunn, M., Wheat, J., Goodwill, S. and Haake, S. (2012) 'Reconstructing 2D planar coordinates using linear and non-linear techniques', in Proceedings of the 30th International Conference on Biomechanics in Sports, pp. 380-383. July 2–6, Melbourne, Australia.

Durkin, J. L. (2008) 'Measurement and estimation of human body segment parameters' in Y. Hong and R. M. Bartlett (eds), *Routledge Handbook of Biomechanics and Human Movement Science*, pp. 197–213. Oxon: Routledge.

Gervais, P., Bedingfield, E. W., Wronko, C., Kollias, I., Marchiori, G., Kuntz, J., Way, N. and Kuiper, D. (1989) 'Kinematic measurement from panned cinematography', *Canadian Journal of Sports Sciences*, 14: 107–111.

Hinrichs, R. N. and McLean, S. P. (1995) 'NLT and extrapolated DLT: 3-D cinematography alternatives for enlarging the volume of calibration', *Journal of Biomechanics*, 28: 1219–1224.

Miller, N. R., Shapiro, R. and McLaughlin, T. M. (1980) 'A technique for obtaining spatial kinematic parameters of segments of biomechanical systems from cinematographic data', *Journal of Biomechanics*, 13: 535–547.

Payton, C. J. and Bartlett, R. M. (eds) (2008) *Biomechanical Evaluation of Movement in Sport and Exercise Sciences – the British Association of Sport and Exercise Sciences*, Guidelines, Oxon: Routledge.

Pribanić, T., Peharec, S. and Medved, V. (2009) 'A comparison between 2D plate calibration and wand calibration for 3D kinematic systems', *Kinesiology*, 41(2): 147–155.

Sanders, R., Psycharakis, S., McCabe, C., Naemi, R., Connaboy, C., Li, S., Scott, G. and Spence, A. (2006) 'Analysis of swimming technique: state of the art: applications and implications', *Portuguese Journal of Sport Sciences*, 6(2): 20–24.

Williams, A. M. and Ericsson, K. A. (2005) 'Some considerations when applying the expert performance approach in sport', *Human Movement Science*, 24: 283–307.

Winter, D. A. (1990) *Biomechanics and Motor Control of Human Movement*, 2nd edn. New York: Wiley.

Yanai, T., Hay, J. G. and Gerot, J. T. (1996) 'Three-dimensional videography of swimming with panning periscopes', *Journal of Biomechanics*, 29: 673–678.

Yeadon, M. R. (1989) 'A method for obtaining three-dimensional data on ski jumping using pan and tilt cameras', *International Journal of Sports Biomechanics*, 5: 238–247.

Yeadon, M. R. and King, M. A. (1999) 'A method for synchronising digitised video data', *Journal of Biomechanics*, 32: 983–986.

Yu, B., Koh, T. J. and Hay, J. G. (1993) 'A panning DLT procedure for three-dimensional videography', *Journal of Biomechanics*, 26: 741–751.

Zhang, Z. (2000) 'A flexible new technique for camera calibration', *IEEE Transactions on Pattern Analysis and Machine Intelligence*, 22(11): 1330–1334.

CHAPTER 5

MOTION ANALYSIS USING ON-LINE SYSTEMS

Clare E. Milner

INTRODUCTION

Biomechanics is about movement, and the objective measurement and recording of three-dimensional human movement is a keystone of the discipline. On-line motion analysis is an essential tool for the study of movement in sport and exercise. As with all tools, motion analysis systems need a skilled and knowledgeable operator to get the most from them. The mark of a good biomechanist is having not only the technical skills to operate a system successfully and collect high quality data, but also the scientific training to use the tools available to further our knowledge of human movement in sport and exercise. In sport and exercise biomechanics, research questions typically have an applied focus, with the aim of furthering our knowledge of elite sports performance or the reduction and prevention of injury. The flip-side of this focus is allowing ourselves to be led by technology, collecting huge amounts of data and trying to find relationships between the many variables involved afterwards, without any clear and logical rationale. In such 'data dredging', biomechanists are seduced by the advanced technology at their disposal, at the expense of scientific rigour.

On-line motion analysis in sport and exercise using optoelectronic motion capture systems is currently being applied to research questions relating to injury or performance in many sports. Many of these studies are investigating injury, trying to elucidate the mechanism of injury or identify the factors that put an individual at increased risk of sustaining an injury. This focus on injury isn't surprising when you consider that remaining injury-free is a fundamental precursor to remain in training and being able to compete successfully. Many research groups have studied running and running injuries, for several reasons, not least that running is a popular recreational exercise for many individuals, not just an elite competitive sport. Running is also associated with a high risk of overuse injury, owing to its repetitive motion. Furthermore, the associations between mechanics and injury are subtle and complex, with little hard evidence of relationships between biomechanical characteristics of runners and injury occurrence being obtained in over 30 years of research. As technology has become more advanced and the research questions posed have become better defined, progress is being made. The biomechanics of athletes in many other sports are being investigated using on-line motion analysis, including rowing, cricket, soccer and golf.

The aim of this chapter is to introduce the key issues that must be considered when designing a motion capture study, from equipment selection to reporting the finished study. There are various competing motion analysis systems on the market, but the basic principles of quality data collection remain constant. Once you have a project in development, and have considered the variables needed to answer your research questions, you will know which data you need to collect. This dictates your hardware and system requirements, which we will consider first. The next stage is designing a data collection protocol that will enable you to collect data efficiently, accurately and precisely. These data are then processed to produce the study variables, which must then be interpreted in relation to the research questions posed and presented in a clear way to illustrate the results of your study for others. Basic principles of good reporting in motion analysis will be considered in the final section.

By the end of this chapter you should be able to identify and address the main quality control issues in on-line motion analysis and have the knowledge to present and interpret the results of a study clearly and meaningfully. Chapters 9 and 10 in this textbook provide the foundations of scientific knowledge to compliment the technical skills developed in this chapter.

EQUIPMENT CONSIDERATIONS

The right equipment is a fundamental consideration for the development of a successful project. Once the variables needed to answer your research question have been identified, you can determine your equipment requirements. In addition to these experimental requirements, several practical considerations must also be borne in mind.

All commercial camera-based on-line motion analysis systems currently rely on some kind of marker tracking. Thus, the initial consideration must be what marker system is most appropriate for the movements that you will study. For example, hard-wired or active marker systems are only appropriate for movements that are contained within a small volume and do not involve multiple twists and turns. The permanent connection of these systems with their markers means that all marker positions are always identified, but may hinder the movement of the athlete. Most on-line motion analysis systems use passive markers to identify the position and orientation of the body in three-dimensional space. These systems rely on the reflection of visible or infra-red light by the highly reflective spherical markers. The reflections are detected by multiple cameras, which record the motion of the body.

A basic hardware consideration is the number of cameras required to track the markers attached to your participants successfully. This will be based on both the number of markers that are attached to the participant and the complexity of the movements being performed. Marker sets and camera positioning will be considered in the next section, but, basically, the greater the number of markers and the more complex the movement, the more cameras will be required to collect good quality data. Minimally, two cameras are needed to enable three-dimensional reconstruction of the location of a marker in space. However, this arrangement would severely restrict both the movements that could be performed and the placement of markers to enable them to be tracked successfully. Adding cameras improves marker tracking and enables more markers to be tracked, up to a point, by increasing the likelihood that at least two cameras will be able to see a marker at each sampling interval. Some camera redundancy is desirable, but as the number of cameras increases, other factors come into play, such as processing time and, in the real world, the cost of the system. The current standard is an eight camera system for sport and exercise biomechanics applications, although many research laboratories operate successfully with six cameras.

Commercially available hardware and software is evolving constantly and ever more advanced equipment is becoming available at less cost. Issues to consider when comparing systems or determining your laboratory's requirements include:

- Type of system; for example, passive or active markers
- Range of sampling frequencies
- Number of cameras that can link to a system
- Maximum camera resolution
- Type of lighting provided; for example, visible or infra-red
- Type and range of lens options
- Minimum useful marker size; for example, in a full body volume
- Real time capability
- Number of analogue channels available to synchronise other hardware; for example, force plate, EMG
- Output file format; for example c3d or ASCII
- Software availability; for example, gait analysis, research
- Service and support options
- Typical price: low, medium or high range

Manufacturers' websites are the best source of up to date information; a list of manufacturers and their websites is provided in Table 5.1.

Table 5.1 On-line systems' manufacturers and their websites.

Company	*Website*
Charnwood Dynamics Ltd	www.codamotion.com
Vicon Motion Systems Ltd	www.vicon.com
Motion Analysis Corp	www.motionanalysis.com
Qualisis AB	www.qualisis.com
BTS Bioengineering Corp	www.btsbioengineering.com
Northern Digital Inc	www.ndigital.com

DATA COLLECTION PROCEDURES

Hardware set-up

There are three key areas that need to be given thorough consideration during protocol development to ensure high quality data are collected: hardware set-up, calibration and marker sets. The stages in hardware set-up are the selection of appropriate camera settings, determination of the optimal capture volume and location of the most suitable camera positions for the chosen volume within the constraints of the laboratory space available.

Check camera lens settings

In general, the camera settings that are manipulated by the biomechanist are the f-stop and the focal length. Some manufacturers offer zoom lenses. F-stop and focal length settings are specific to the size of lens being used and recommended settings for a particular system will be provided by the manufacturer. The f-stop is related to the amount of light allowed to pass through the lens, with a 'larger' f-stop (f-16 rather than f-2, for example) letting less light through the lens. Generally, the focal length is set to infinity to get the maximum depth of field from the camera view.

Determine the location and dimensions of the capture volume

The size of the capture volume is a key consideration, since it affects the resolution of the system and, therefore, the precision with which position data can be recorded (see Chapter 9). The chosen capture volume is a compromise between the need to accommodate the movement being studied and maximising the resolution of the system by using the smallest volume possible. The volume must be sufficiently large to accommodate the movement in question fully for all expected sizes of participant. It is important to remember that this may need to include, for example, the hand of an outstretched arm in throwing sports, not just the torso. However, the volume only needs to cover the portion of the movement that is being studied. For example, a complete running stride, from foot strike of one foot to the next foot strike of the same foot may be two metres long. However, if the research question relates only to the stance phase of running, the length of the capture volume in the direction of running could be reduced from around 2.5 m, which would be needed to ensure that a complete stride occurs within the volume, to around 1 m.

Although it may appear simplest to use a large capture volume to ensure that no movement data are lost, this will adversely affect the resolution of the system, resulting in a loss of precision in your data. The field of view of each camera is composed of a fixed number of pixels, for example 887 rows by 1120 columns (1 Megapixel), although high resolution camera systems have resolutions up to 3456 rows by 4704 columns of pixels (16 Megapixels). Regardless of the specific resolution, the field of view of the camera is represented by a fixed number of pixels. A pixel is the minimum precision to which a marker in the camera view can be located. If the chosen capture volume is large, each pixel represents a relatively larger part of the 'real world' than it would for a smaller capture volume. For example, if the width of the capture volume was 5 m, each of 1280 pixels in the width of the camera view would represent 5000/1280 mm or 3.9 mm of that width. However, if the width of the capture volume could be reduced to 1 m, each pixel would represent 1000/1280 mm or 0.8 mm. Resolution is also affected by camera position in relation to the capture volume.

Place cameras appropriately around the chosen capture volume

In setting up a motion analysis laboratory, maximum flexibility in camera placement should be maintained. Wall-mounting of cameras is often used to keep the cameras out of harm's way and reduce the risk of them being knocked over or damaged during other laboratory activities. If cameras are to be wall-mounted, lighting arms provide some flexibility in camera position in all three planes. Depending on the size of the laboratory and the floor space available, this may be the most feasible option. Maximum flexibility is maintained by the use of camera tripods to position the cameras; however, these require a lot of floor space and are more vulnerable to accidental contact, which then requires at least the recalibration of the system and may knock the camera to the floor and damage it. Other camera mounting arrangements have included tracks around the walls of a laboratory, along which the cameras can be moved horizontally and vertically. The disadvantage of this arrangement is that the cameras cannot be brought any closer to the capture volume to maximise system resolution. A variation on the tripod method is the use of vertical posts mounted on the laboratory floor, with or without overhead horizontal bars between them. These require considerably less floor space than tripods, but must be attached to the floor to remain upright and stable, reducing the flexibility of the set-up. Addition of the horizontal bars increases the camera positioning flexibility in this arrangement. In the typical motion analysis laboratory, which is used for multiple concurrent projects, the flexibility gained by the use of tripods is often preferable, as it enables cameras to be positioned around the most appropriately sized capture volume for each study.

Ensure 'dead space' in each camera's field of view is minimised

The most appropriate camera position is that which minimises the 'dead space' surrounding the chosen capture volume in the camera's field of view. Dead space is that part of the camera's field of view that will not contribute to the data collection because it lies outside the capture volume. The presence of dead space reduces the resolution of the system because it effectively reduces the number of pixels that are available to represent the capture volume. A small dead space is usually unavoidable, given the fixed aspect ratio of the camera view and the shape of the capture volume, particularly when the volume is viewed from a position that is not orthogonal to it. However, dead space in the camera view should be minimised by ensuring that the camera is close enough to the volume for its field of view to be filled by the volume. If your cameras have the traditional rectangular field of view, it may be necessary to rotate them through 90° to minimise the dead space around the capture volume. Most systems have an option that enables you to rotate the apparent view of the camera so that the image seen on the computer screen appears in its normal orientation, despite the rotation of the camera.

Typically, the cameras will be equally spaced around a cuboid capture volume. This arrangement maximises the separation between adjacent cameras and provides maximum coverage of the capture volume, reducing the likelihood of marker dropout. This happens when a marker is not seen by at least two cameras in a sample and, thus, its position cannot be reconstructed. In reality, laboratories are often not an ideal shape and camera position options may be further reduced by the presence of other equipment. Additionally, different sports may require unusually shaped capture volumes. For example, in studying a tennis serve, particularly if the path of the head of the racket is of interest, a very tall, but relatively narrow volume will be required to accommodate the height of an arm plus racket stretched overhead. A good starting point is still to space the cameras equally around the volume and then fine tune the individual positions to minimise dead space while ensuring that the entire volume can still be seen by the camera.

All of these hardware issues are interrelated. The choice of camera lens dictates how much of the laboratory space the field of view of the camera will see. In turn, this is related to the position of the camera in relation to the chosen capture volume to ensure that dead space is minimised. For example, a wide-angle lens will need to be closer to the capture volume than a normal lens to reduce the dead space around the capture volume in the camera's field of view by a similar amount. Lens choice should reflect the experimental set-up.

Select an appropriate sampling rate for your set-up and experiment

Other considerations when setting up your system for data collection include the sampling rate, or frequency, and synchronisation with other equipment. In some cases, the sampling rate can be specified, with higher sampling rates being associated with decreased resolution in a trade-off associated with the processing power of the hardware. For example, cameras that have full resolution at 240 Hz (samples per second) may be able to capture data at 480 Hz, but only at 50 per cent of the spatial resolution. That is, there is a trade-off between temporal resolution and spatial resolution. Again, the most appropriate option for your study should be selected. For example, gait analysis data are typically collected at 120 Hz, whereas jump landing data are typically collected at 240 Hz. If the capture volume is small, the loss of spatial resolution may be acceptable to allow a higher temporal resolution to capture a high-speed movement. The loss of spatial resolution associated with a large capture volume may be unacceptable and a lower temporal resolution must be chosen. Most human movements have a relatively low frequency, but where high speeds of movement and rapid changes of direction are involved, movement information may be lost if the sampling frequency is too low (see Chapter 9 for more detail on sampling time-series data).

Synchronise other hardware at an appropriate sampling frequency

Most motion analysis systems can be synchronised with other biomechanical hardware through a series of analogue input channels. This allows force platform, EMG or other analogue data to be time-synchronised with the kinematic data collection of the motion capture system. Other hardware typically samples at a higher frequency than the motion capture system. To enable the data sets to be time-synchronised, the higher frequency should be a multiple of the motion capture frequency.

Calibration

Calibrating the motion capture system enables the image coordinates on each individual camera view to be converted to the real world three-dimensional coordinates of each marker. For most systems this is a two-stage process with an initial dynamic, or wand, calibration followed by a static calibration.

Adjust camera sensitivity

For the calibration to succeed, camera sensitivity must be checked. The camera sensitivity changes the reflected light intensity at which the camera will register a reflection and try to reconstruct it as a marker. If the sensitivity is set too high, lots of stray reflections will be picked up in addition to those from the actual markers. If too low, the markers themselves will not be recognised. It is common practice to place the calibration wand within the capture volume and use it to adjust camera sensitivity. If the calibration wand is placed at the origin of the laboratory coordinate system, as is also necessary for the static calibration, one can also confirm that all cameras can view the laboratory origin. Camera sensitivity should be adjusted for each camera individually until all calibration wand markers are clearly seen in each field of view. Sensitivity set too high will result in 'fuzzy' marker outlines, whilst set too low will result in small markers that may not be reconstructed.

Check for stray reflections in each camera's field of view

When the sensitivity of each camera has been adjusted, the individual camera views should be checked for additional reflective objects, for example, other cameras. It is good practice, and a necessity for some types of software, to ensure that camera views are adjusted to avoid stray reflections being picked up. If a reflection is picked up by more than one camera, it will be reconstructed as a marker and the static calibration will fail. However, most software enables those parts of the camera view that contain an unwanted object to be blocked out, or masked, but this part of the view is also blocked during data collection. The more cameras in a motion capture setup, the greater the likelihood of a camera being picked up in another camera's field of view and being incorrectly identified as a marker. In this case, software masking of stray camera reflections will maximise the performance of the setup.

Perform dynamic calibration

The dynamic calibration is conducted to register the cameras to the whole of the capture volume. The wand calibration is basically the direct measurement of an object of known dimensions made by all the cameras throughout the entire capture volume. Each system manufacturer has its preferred style of wand calibration, but the fundamental requirements are that the whole volume is covered and that the orientation of the wand is varied to ensure accurate calibration in all three cardinal planes. Care should be taken to cover the whole of the capture volume. Straying a little outside the capture volume boundaries in each direction will help to ensure that the whole volume is calibrated.

Determine the origin and orientation of the laboratory-fixed global axes

The static calibration uses the calibration wand with five markers mounted in known locations (Figure 5.1). The wand defines the location of the origin and the orientation of the laboratory reference frame. Typically, the corner of a force platform is chosen as the origin. This improves consistency across data collections as the force platform is rigidly mounted in the laboratory and provides a consistent orientation for the laboratory reference axes. The orientation of the laboratory reference frame has no effect on the joint angles calculated from the three-dimensional coordinate data, but standard definitions have been suggested to make communication between laboratories easier. The International Society of Biomechanics (ISB) has suggested aligning the X axis with the direction of progression, the Y axis vertically and the Z axis medio-laterally to create a right-handed orthogonal coordinate system (Wu and Cavanagh, 1995). This orientation was suggested as it is a simple extension from sagittal plane two-dimensional analyses that have X oriented horizontally and Y vertically. Extension into the third dimension simply requires the addition of a medio-lateral axis, the Z axis. A typical engineering definition of the laboratory coordinate system places X in the direction of progression, Y medio-lateral and Z vertical. It should be noted that the Z axis in the ISB definition is equivalent to – Y in the engineering version, as both follow the right-hand rule.

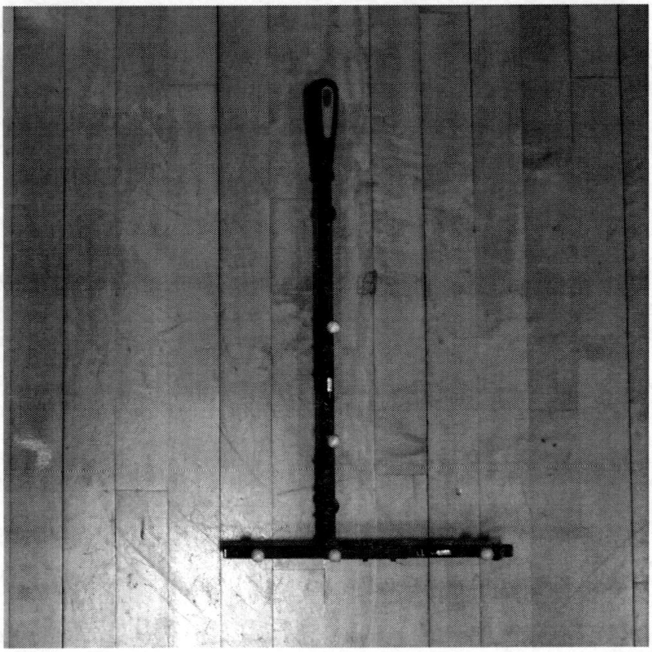

Figure 5.1 The calibration wand used in the calibration of a motion capture system.

Perform static calibration

The static calibration is the second stage of the two-stage calibration process. It is important to place the calibration wand in the correct orientation with respect to the laboratory, and any additional hardware such as force platforms, if the global axes are to be used as a reference for any of your variables. This is of key importance if body positions will be normalised with respect to their pose in the laboratory reference system, or absolute segment positions will be reported, for example toe-out angle during running. The only markers that can be present in the camera views during calibration are those on the wand.

Determine an acceptable residual for your set-up

A good calibration is fundamental to the collection of high quality data. The position of the markers in three-dimensional space can only be located to the accuracy to which the system was calibrated. At the end of the calibration process, the data capture software typically provides an indication of the precision that can be achieved based on the calibration performed. Each camera has a residual associated with it. This number, in the units in which the system was calibrated, provides an indication of the precision to which the marker position can be located. For example, a residual of 1.0 mm indicates that a marker's position in space will be located to within 1 mm of its true position. Larger capture volumes will have higher residuals for the same quality of calibration owing to the resolution of the system being based on a fixed number of pixels in each camera view (see preceding Hardware set-up section).

Ensure a valid calibration is maintained throughout the data collection session

Calibration should be done before every data collection session, even if the cameras are wall-mounted and do not appear to have moved between sessions. Calibration is a sensitive procedure and even a small vibration of a camera can move it sufficiently to reduce the quality of the reconstruction of marker location. If a camera is knocked or moved during a data collection session, the system must be recalibrated. Every effort should be taken to avoid this, such as asking participants to take care not to touch the cameras and having as few people as possible moving about the laboratory during data collection. However, the calibration procedure is easy and takes only a few minutes. If in doubt, it is always worth recalibrating to be sure that you are maintaining the quality of your data. A good indicator that the quality of your calibration has been reduced is the appearance of ghost markers in your capture volume. Ghost markers are markers that are reconstructed by the software based on the information from the cameras and the calibration parameters, but that do not actually exist. Other reasons to recalibrate include changing the units of measurement or the orientation of the laboratory coordinate system, although in practice these changes would be unlikely to occur during a data collection session.

Marker sets

A marker set is simply a configuration of markers that meets the selection criteria defined originally by Cappozzo et al. 1995. This foundational paper is a classic which is still cited widely today in the procedures of motion analysis studies.

A minimum of three non-collinear markers is required per rigid segment for three-dimensional analysis

Three non-collinear markers are required on a segment to define its position and orientation in three-dimensional space. Two markers define a line and can provide information about two-dimensional movement but cannot indicate whether rotation about that line is occurring. A third non-collinear marker, which does not lie on the line, is needed to define this axial rotation of the segment.

Each marker must be seen by at least two cameras at every instant during data recording

This is the minimum requirement for reconstruction of the marker position in three-dimensional space, although reconstruction accuracy may be improved when some redundancy is introduced and more cameras can see the markers. Furthermore, the inter-marker distances must be sufficient that the system can differentiate between individual markers during marker reconstruction. If markers are placed too close together, the system will be unable to identify where one marker begins and the other ends. In this case, only one marker will be reconstructed instead of the two. Additionally, the further apart segment markers can be placed, the smaller the impact of marker position reconstruction error on the joint angles that are calculated from the raw coordinate data.

Minimise movement between the markers and the underlying bone

The aim of on-line motion analysis is to determine the movement of the underlying bone structures by recording the movement of markers mounted on the skin's surface. The discrepancy between the movement of the marker and the actual movement of the skeleton is known as 'skin movement artefact'. The gold standard for marker mounting is to attach markers directly to the bones of interest using bone pins. Although this has been done in a few studies, it is obviously impractical in most cases as it is invasive. However, comparison studies between bone pin and skin-mounted markers provide an indication of the magnitude and type of error associated with skin-marker measurement. One study reported errors due to skin-marker movement at the knee during walking may be up to 10 per cent of the flexion – extension range of motion, 50 per cent of the abduction – adduction range of motion and 100 per cent of the internal – external rotation range of motion (Cappozzo et al., 1996). Those authors noted that minimising skin movement artefact must be the main criterion in marker set design (Cappozzo et al., 1996). Errors due to skin movement can be reduced by choosing the most appropriate marker attachment methods for your study.

Marker set selection and development

Determine the type of marker set that is most appropriate for your study

The choice of an appropriate marker set is critical to ensuring that all of your required variables can be calculated from the raw data collected. It also plays a key role in the quality of your data set, related to skin movement artefact and the introduction of noise into the data. Choosing an appropriate marker set and mounting it accurately on the skin is essential to obtaining data that are representative of the movement being studied. There are three main options when determining the marker set to use in your study (Figure 5.2). The first is to use one of the standard clinical gait analysis marker sets to answer your question. Alternatively, you can design your own marker set with markers attached to anatomical points of your choice. Finally, you could use markers attached to rigid thermoplastic clusters that are then attached to the participant. Each of these methods has advantages and disadvantages.

(a)

(b)

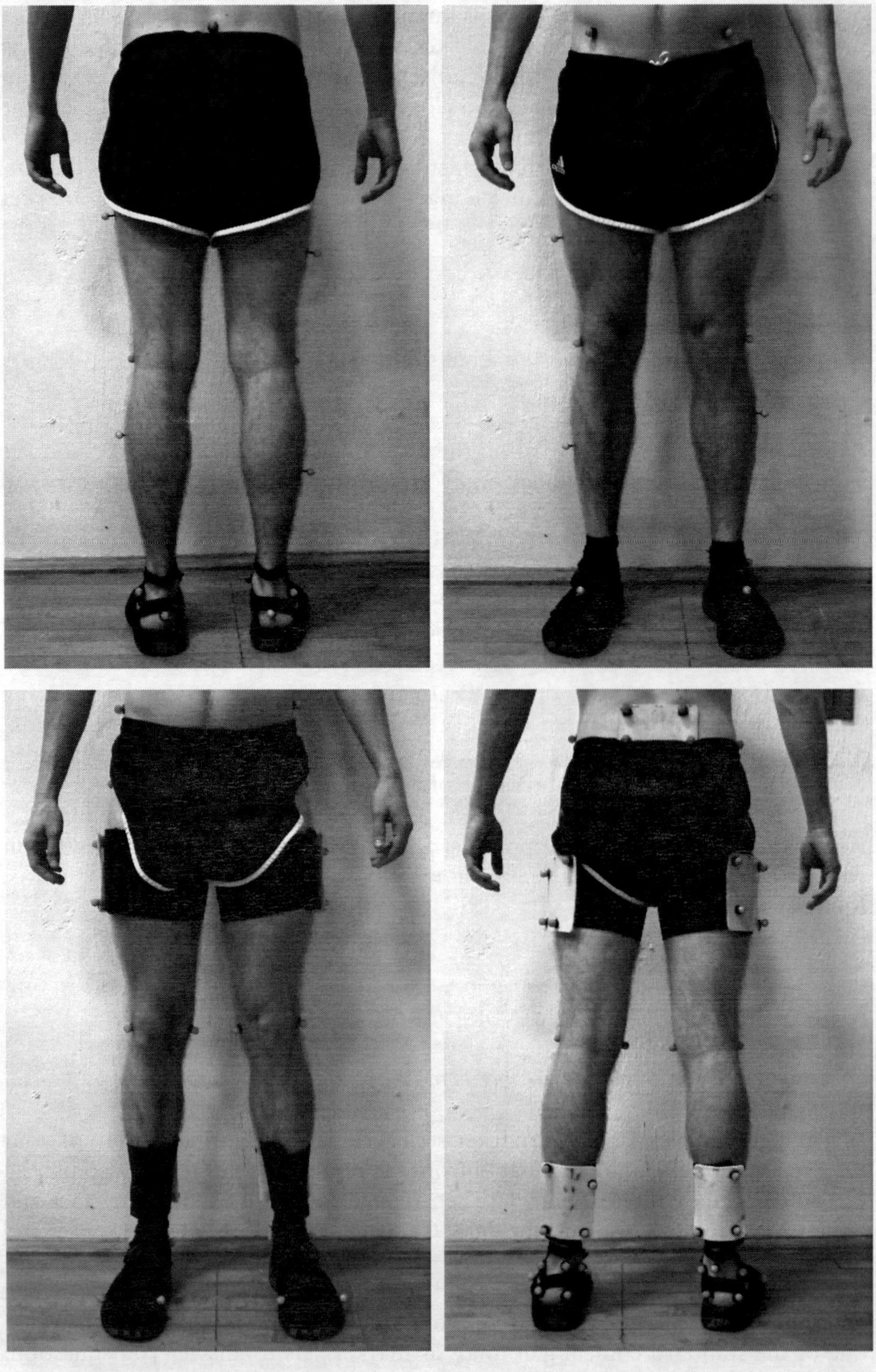

Figure 5.2 Marker sets used in on-line motion analysis: (a) standard clinical gait analysis marker set; (b) cluster-based marker set.

Use the standard clinical gait analysis marker set

The advantages of a standard clinical gait analysis marker set, for example Plug-in-Gait (Davis et al., 1991 and Kadaba et al., 1990), are that it has been tried and tested in many laboratories throughout the world and has been used in very many studies over the past 20 years. The main manufacturers also provide software that will automatically process raw data obtained using such marker sets through to time-normalised mean graphs for each variable. However, such marker sets have several serious limitations, related to the small number of markers used. The most serious limitations tend to be at the foot, which usually has only two markers on it. Consequently, the multi-segmented foot is represented simply as a line, making true three-dimensional rotation information about this segment, relative to the leg, impossible. Additionally, the shank often has only two markers on it (the only marker at the knee joint is located on the femoral condyle). Furthermore, since such marker sets were designed originally for lower extremity gait analysis during walking, applying them to sports movements may not be valid and may reduce their effectiveness, both in terms of markers being obscured from camera views and also in minimising marker movement. For example, the placement of markers on wands to improve the estimation of transverse plane rotations may result in unacceptably high marker vibration at impact during higher speed activities. However, this source of error can easily be reduced by placing these markers directly on the skin.

Design a custom marker set

An alternative is to design a custom marker set for your study that follows the rule of three markers per rigid segment; necessary for true three-dimensional motion capture. This enables some of the limitations highlighted previously to be overcome. However, a good knowledge of surface anatomy is needed to determine the most appropriate marker locations that both minimise skin movement artefact and enable joint centres to be located. Typically, two markers are needed to define the relative location of a joint centre and the third marker is placed somewhere in the middle of the segment away from the line between the two anatomical markers to enable the three-dimensional position and orientation of the segment to be recorded. Additional markers may be placed on a segment to ensure that at least three are visible in every frame of data collected. This is a consideration for complex sports movements that involve twisting and bending of the body that may obscure certain markers.

Develop and use a cluster-based marker set

The third option is the use of marker clusters mounted to thermoplastic shells that are placed on the middle region of the segment and have a known relationship to the anatomical points required for data processing. This method allows more flexibility in camera placement as the clusters can be positioned on the segment in the orientation that makes them most visible to the cameras. This is particularly useful if only a few cameras are available or the laboratory environment restricts camera placement options. Use of marker clusters on the shank segment was shown to better reflect motion of the underlying bone compared to individual markers placed directly on the skin in a bone-pin study (Manal et al., 2000). However, this method necessitates an extra step – the collection of a static trial. Static trials are often used with the other types of marker sets to provide a zero reference position for joint angles and to define joint centre locations. However, with marker clusters one or a series of static trials is required to locate the anatomical points relative to the cluster. If a single static trial is used, markers are placed on all of the anatomical points and their relationships to the cluster calculated during data processing. An alternative method (Cappozzo et al., 1995) uses a pointer to locate each anatomical point. This method requires a separate static trial for each anatomical point, but does not require that all of the anatomical markers are visible simultaneously, unlike the single trial method. Therefore, it can be used in those setups that have restricted camera placement options.

Introduce asymmetry into your marker set to optimise marker tracking

Asymmetry can be introduced in your marker set simply by adding an additional marker to one side of the body. Additionally, non-anatomical markers on each segment can be varied in placement between the right and left sides. The asymmetry enables the software algorithms, which use the distances and angles between markers to identify them, to work more effectively.

Consider marker size

Marker size is a compromise between having a marker large enough to be seen by several cameras and to cover multiple pixels, which improves the accuracy of the location of its position, but not so large that it interferes with movement of the athlete or overlaps with other markers in the camera view. The standard clinical gait analysis marker set uses 25 mm markers, but high-resolution cameras can be used successfully with 5 mm markers.

Use an appropriate method of marker attachment

Mounting the markers on the skin must be fast and easy and the attachment must remain secure for the duration of data collection. Various methods are used for marker attachment, although directly on the skin is preferable. Markers attached to clothing have another layer between them and the bone, introducing the possibility of further movement artefact. Double-sided tape, such as toupee tape and electrocardiography (ECG) discs are used to attach the markers. If these are insufficient, adhesive sprays may be used to provide additional adhesion, for example if the participant will sweat during the data collection. In some instances tight-fitting Lycra clothing is used and markers are attached using Velcro. This method is quick and avoids any discomfort for the participant associated with pulling adhesive tape off the skin, but introduces the extra layer of movement between the marker and the bone. The most effective method of marker cluster shell attachment appears to be attachment to a neoprene underwrap secured with hook and loop tape, as shown on the shank and thigh in Figure 5.2(b).

Take great care with marker placement

It must also be emphasised that the marker set is only as good as the person who applies it to the participant. If the markers used in the determination of joint centre locations are misplaced relative to the underlying bony anatomy, the joint centres used in data processing and analysis will not be coincident with the actual joint centres, so cross-talk will be introduced into the kinematic data. Consistent marker placement between investigators and between days is notoriously difficult to achieve. It is recommended that only one person places markers on all participants in a study to remove inter-investigator variability as a source of error.

PROCESSING, ANALYSING AND PRESENTING MOTION ANALYSIS DATA

Data processing

The first stage in data processing is cleaning the data, which means checking that the markers are identified correctly in every frame, joining any broken trajectories and deleting unnamed markers. Current on-line motion capture software removes the need for manual identification of markers in each frame of data collected. In the past, this onerous task was the rate-limiting factor in processing human movement data. Nowadays, marker identification either occurs in real-time or takes place in a matter of seconds after data collection. However, the software is not fool proof and each trial collected must be checked to ensure that markers have been identified correctly. This is simple pattern recognition as the template used to join the markers can be recognised and deviation from the correct template can be readily seen by eye, especially if the template is colour coded. The additional work needed at this stage can be minimised by designing a robust marker set, with planned asymmetries, and an associated template before data collection.

Second, if a good camera set-up and appropriate choice of marker set is made initially, the number of broken trajectories and size of the breaks should be minimal. There are two schools of thought on joining broken trajectories. One is that any line joining is unacceptable and the hardware set-up should be refined and new trials recorded until complete trajectories are obtained. The other is that small breaks in marker trajectories that are not at peaks or changes in direction of the curve are acceptable for around five consecutive data samples. Most software offers quintic spline curve-fitting techniques, which adequately reproduce the curve provided that the curve is only undergoing minor changes during the period when data are missing. This makes practical sense in that whole trials do not need to be rejected as a result of a minor dropout of a single marker for a few samples, but data are rejected if multiple markers are lost or the period of marker dropout is significant.

Third, unnamed markers need to be deleted from the trial before the data is processed. These unnamed markers are the ghost markers referred to earlier and do not contribute any information about the movement of the participant. Before deleting these markers, you should make sure that all of the body markers are identified correctly and none has been switched with a ghost marker by the template recognition algorithm of the software. Additionally, you should ensure that apparently broken trajectories are not simply points where the algorithm has failed to recognise a body marker and it has become temporarily unnamed. When you are satisfied that any unnamed markers are, indeed, ghost markers, they should be deleted from the trial so that the data processing software functions properly.

Filtering is another important consideration with human movement data. Since it is covered in detail in Chapter 9 of this book, only basic considerations will be raised here. Noise due to skin marker movement is generally of high frequency and human movement occurs at low frequencies; therefore, the two can be separated using a low-pass filter. Such a filter separates the components of the time-displacement curve of a marker based on whether the components oscillate above or below a chosen cut-off frequency, and discards all components above the cut-off. The cut-off frequency used can have a significant effect on the shape of the marker trajectory: too low and the curve becomes oversmoothed – the peaks are flattened; too high and the curve remains noisy. Cut-off frequencies can be determined using specific algorithms, plotting a frequency distribution curve or by trial and error sampling of different cut-off frequencies and observing their effect on the data. As an indication of cut-off frequencies used for different movements, walking data are typically filtered at 6 Hz, running at 8 Hz to 12 Hz.

Once the raw data have been prepared in this way, they can be analysed to obtain the kinematic variables needed to answer your research question.

Data analysis

The methods of data analysis used in your study depend on the type of system you have and the marker set that you used. If you use the standard gait analysis marker set, most systems provide software that processes the raw data automatically and provides plots of all the variables as output. This method is simple to use, but does not give the biomechanist any control over the calculation methods. If you choose to use a custom marker set you may be able to use the research software provided by the system manufacturer. This allows you to define the segment coordinate systems and other components of your data analysis from options written into the software. This additional flexibility provides more control over the data analysis stage of your study, but relies on your technical knowledge being sufficient to make the correct decisions about segment coordinate axis definitions, joint coordinate system calculations and orders of rotation. For complete control over the data analysis process, the final option is to write your own software using a computer or engineering programming language. With this method you can be confident that all of the calculations are exactly as you intend and you have complete flexibility in your choice of data collection method. However, writing and validating such code for three-dimensional kinematic analysis is a major task requiring excellent technical and mathematical knowledge and a significant time commitment; it should not be undertaken lightly.

Regardless of the method chosen, the end result of data analysis is the extraction of three-dimensional kinematic data from the raw marker position data collected during the movement trials. The basic output is angular variables, or their derivatives, against time for each trial. These data may then be further manipulated or reduced to aid data interpretation.

Data interpretation and presentation

There are many options for interpreting and presenting motion analysis data. To aid the interpretation of reduced data – discrete values such as peaks – it is useful to present time-normalised mean curves to enable qualitative interpretation of the data. It should always be remembered that the average curve smooths the peaks of the individual trials as a result of inter-trial temporal differences. The most appropriate way of presenting these data will depend on your project, and the variability between trials (Figure 5.3). If all trials are very similar, a single mean curve may be most appropriate. However, an indication of the variability between trials is usually helpful, and a mean ± 1 standard deviation curves will often be more informative. In other cases, it may be better to plot all of the individual curves on the same graph, to give the reader all of the information about inter-trial variability. Mean values for the key variables are reported alongside these plots.

Data reduction means extracting key variables from the continuous data obtained during data processing. Typical variables are peaks of angular displacement or velocity, or angles at defined instants. For example, peak rear-foot eversion and inversion angle at foot strike are discrete variables that are often considered in research questions about running injuries. The key decision to make when reducing data is whether to extract the discrete data points from the time-normalised mean curve, or from the individual trial curves and then calculate a mean value for the variable in question. For times that are clearly defined by an event in the time domain, the output will be the same. For example, rear-foot angle at foot-strike is unaffected because it defines the same data point on each individual curve as on the mean curve. However, extracting data from the individual curves is more appropriate for peak values because it preserves the time domain variability information. The mean curve tends to reduce the apparent peak values when they occur at different times.

Depending on the research question asked, other types of analysis may be appropriate. For example, much current research is concerned with the coordination between segments and how coordination variability may affect injury

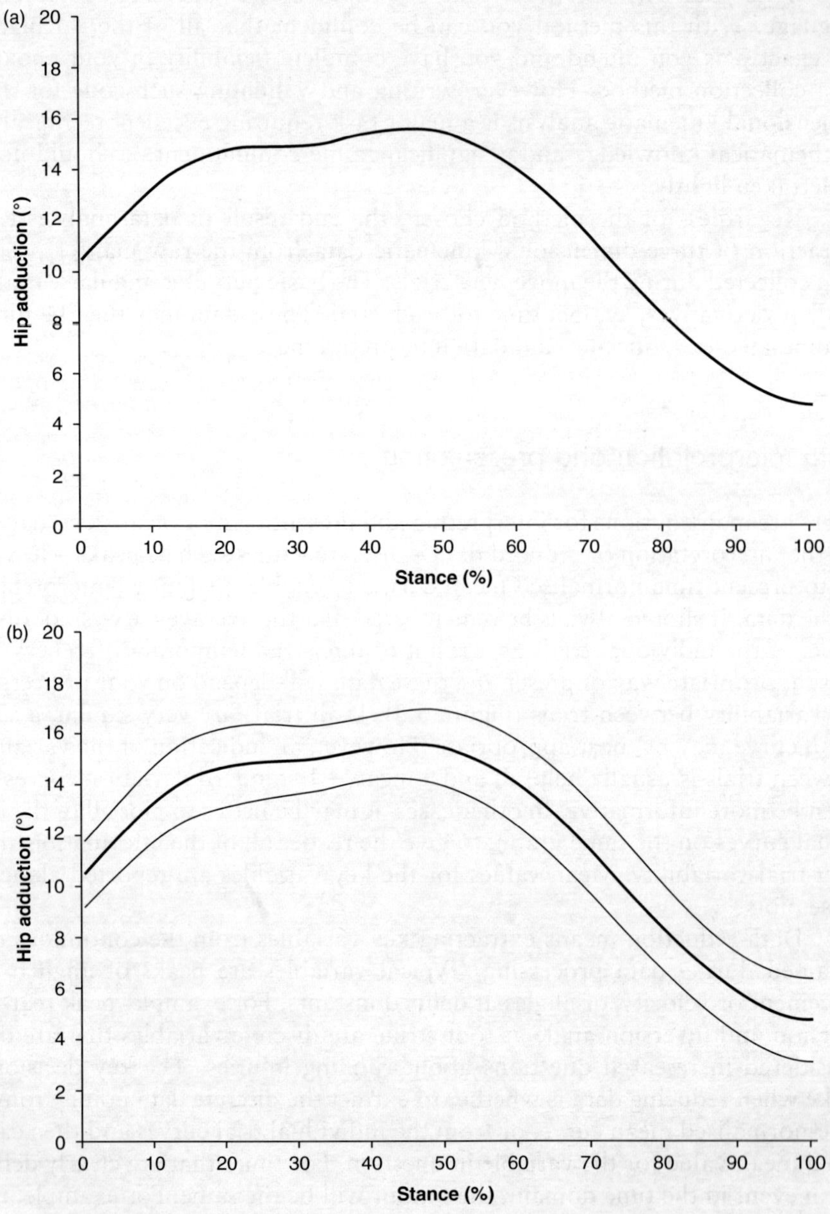

Figure 5.3 Different ways of presenting the same multiple-trial time-normalised kinematic data: (a) mean curve; (b) mean ± 1 standard deviation curves; (c) all individual curves. The example shown is hip adduction angle during running.

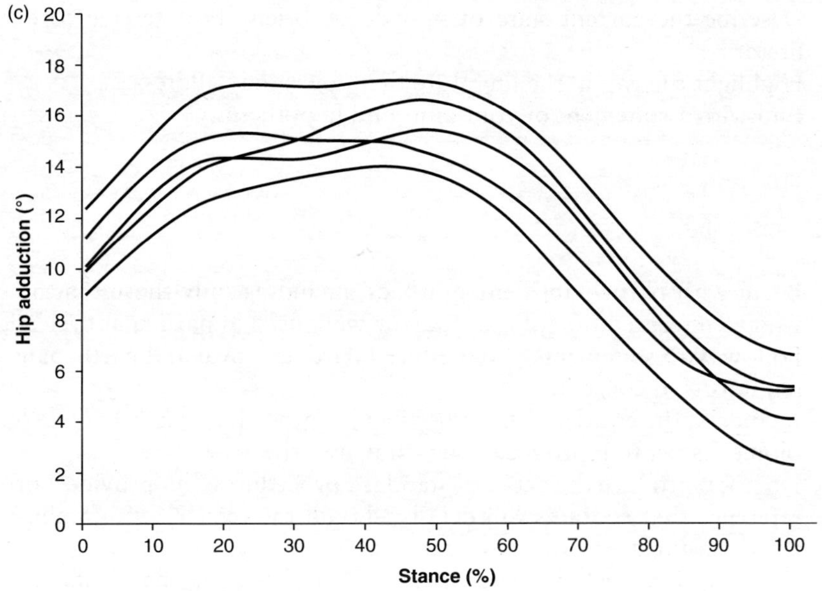

Figure 5.3 Continued

risk. Techniques used to quantify and assess coordination and variability are addressed in Chapters 3 and 10 of this book, respectively.

The common thread in current analyses of human movement data is the extraction of discrete variables from continuous time-series data. The discrete variables are used to answer your research questions statistically. To maintain the statistical power of your study, it is preferable to identify a few key variables that will contribute the most to your hypotheses and present other associated variables, continuous and discrete, descriptively to aid in the interpretation of your statistical results.

REPORTING A MOTION ANALYSIS STUDY

This section contains general guidelines for reporting a motion capture study. The detail required in specific sections will vary depending on the purpose of the report: whether it is a report to a research group or commissioned research for a consultancy client. For example, the client that commissioned a piece of research may only be interested in broad-brush detail of your methods, but may require in depth explanation of your results in non-technical layman's terms.

Introduction

- Introduce the general area in broad terms.
- Focus on the specific topic of your study.

- Describe the current state of knowledge briefly by referring to existing literature.
- Highlight the gap in the literature that your study addresses.
- End with a statement of your aims and hypotheses.

Methods

- Begin with participant demographics, including only those that are relevant to interpretation of the study or were used in participant selection.
- Follow with statements about ethics board approval and participant written informed consent.
- Include sufficient detail of experimental procedure for a knowledgeable biomechanist to reproduce your study exactly.
- Do not reproduce the detail of standard procedures, but provide a primary reference that contains full details, should the reader be unfamiliar with the procedures.
- If appropriate, include a section on statistical analysis that contains details of the tests used and the dependent and independent variables used to test your hypotheses.

Results

- Part of the skill of a biomechanical investigator is to be able to tease out those data that are relevant to their research questions and use these to construct a focused and informative results section.
- Include only those data that address the aims of your study. Including every result that your data processing software spilled out does not impress the reader with how much work you did. Instead, it forces them to conclude that you did not know what question your study was trying to answer.
- The results section should contain details of the variables that were tested statistically, and the results of the statistical tests done. It helps the reader to interpret your results if the actual P values obtained are reported, rather than whether or not a particular level of significance was reached.
- Descriptive statistics of those secondary variables that were not related directly to your hypotheses, but aid in the interpretation of the primary variables, should also be included.
- Figures and tables should be included as necessary for clarity, but should not replicate the same information.
- Text in this section should summarise the results and draw attention to relevant details contained within figures and tables.
- Remember that the purpose of this section is to present the results of your study: their interpretation occurs in the Discussion section.

Discussion

- The results presented are interpreted with reference to the literature that was included in the Introduction.
- Do not introduce new concepts or literature in this section. The Discussion ties together all of the information that has been presented in the Introduction, Methods and Results sections. Any existing literature that you want to refer to should be included in your Introduction and any results you want to discuss should be contained within the Results section.
- Comment on whether your initial hypotheses were supported or rejected by the results obtained and discuss the implications of this.
- Summarise what has been learned from your study and draw conclusions about your research questions, if this is appropriate.
- Be careful not to make grand sweeping statements based on your study. Increasing knowledge and understanding of a biomechanical problem is a stepwise process. Acknowledge the contribution your study makes, but avoid extrapolating beyond its parameters.
- Be sure to acknowledge any limitations of your study and how these might affect the interpretation of your results.
- Discuss whether the aims of the study set out in the Introduction have been achieved and, if not, comment on why not and what might be done in the future to address this.
- If the study is part of a bigger picture, it may be appropriate to refer to future plans to build on the knowledge gained and address further relevant questions in the research topic.

References

- Include a complete list of all the literature that has been referred to in the body of your report. It is important that the list is accurate to enable the reader to follow up on aspects that are of particular interest.
- Literature that you want to highlight to the reader that has not been referred to directly in the report should be put in a bibliography at the end of the report. Note that a bibliography is not usually included in scientific reports, but may be useful in a report to a consultancy client to enable them to learn more about specific areas.

The most important consideration when compiling your report is that it follows a logical order, from the development of a research question through data that relates to that question and, finally, ties together existing knowledge and new results from your study to answer the research question and add another piece of understanding to the puzzle of human movement in sport and exercise.

Having reached the end of this chapter, you should now be familiar with all of the main quality control and technical issues related to on-line motion analysis as a data collection tool in sport and exercise biomechanics. Every project is different and has specific requirements and pitfalls. Preparation is the key to success and time spent addressing the issues raised at the outset of the study will help to prevent problems later, after the participants have made their contributions.

ACKNOWLEDGEMENT

I would like to thank Trey Brindle for kindly agreeing to act as the model for Figure 5.2.

REFERENCES

Cappozzo, A., Catini, F., Della Croce, U. and Leardini, A. (1995) 'Position and orientation in space of bones during movement: anatomical frame definition and determination', *Clinical Biomechanics*, 10: 171–178.

Cappozzo, A., Catini, F., Leardini, A., Benedetti, M. G. and Della Croce, U. (1996) 'Position and orientation in space of bones during movement: experimental artefacts', *Clinical Biomechanics*, 11: 90–100.

Davis, R. B., Ounpuu, S., Tyburski, D. and Gage, J. R. (1991) 'A gait analysis data collection and reduction technique', *Human Movement Science*, 10: 575–587.

Kadaba, M. P., Ramakrishnan, H. K. and Wootten, M. E. (1990) 'Measurement of lower extremity kinematics during level walking', *Journal of Orthopaedic Research*, 8: 383–392.

Manal, K., McClay, I., Stanhope, S., Richards, J. and Galinat, B. (2000) 'Comparison of surface mounted markers and attachment methods in estimating tibial rotations during walking: an in vivo study', *Gait and Posture*, 11: 38–45.

Wu, G. and Cavanagh, P. R. (1995) 'ISB recommendations for standardization in the reporting of kinematic data', *Journal of Biomechanics*, 28: 1257–1261.

CHAPTER 6

MEASUREMENT OF EXTERNAL FORCES

Nachiappan Chockalingam and Aoife Healy

INTRODUCTION

Quantitative assessment of external forces is important to understand human movement. These external forces are applied when there is contact between the individual and the immediate environment, either through direct contact or through indirect contact via sports equipment or supporting devices such as prosthetics, orthotics or wheelchairs. This contact could be experienced by any part of the body depending on the type of movement and the type of devices or equipment used. In some instances, it will be useful to measure the resultant force or, in other instances, it will be more relevant to measure localised forces. When one uses appropriate technology, the localised force can lead to the assessment of pressure. Force and pressure measurement forms a fundamental part of understanding human movement in the context of physical activity and sport.

Information on the forces acting during physical activity can help provide a greater understanding of the motion of the body to achieve the desired overall movement. Such information can assist in designing and delivering comprehensive coaching and/or exercise interventions, understanding injury mechanisms and the formulation of effective clinical management and rehabilitation.

Whilst there are several technologies and techniques available for force and pressure measurement in sport and physical activity, the purpose of this chapter is to provide an introduction to the main instrumentation and procedures in this area. It is not the intention of the authors to provide a comprehensive list of all available sensors or technologies. Rather, we will provide general principles illustrated with specific examples regarding the use of commercially available force platforms to measure ground reaction forces, as well as plantar pressure measurement systems.

MEASUREMENT OF GROUND REACTION FORCES

A number of forces are present during all sport and exercise activities. Some of these forces are internal, that is, they act within the body, and often occur as a result of muscle activity. Other forces are external, that is, forces acting on the body due to the interaction of the individual with their immediate external environment. Within a sporting context, whilst there can be several external forces acting, including air resistance, weight and buoyancy, the major external force present during terrestrial activities, such as running and jumping, is the ground reaction force (GRF). This is the external force acting on the body in reaction to its contact with the ground (or any external surface). The GRF is conventionally resolved into three orthogonal components. For a level surface, the components are: vertical, which is normal to the surface, and two shear (usually frictional) components, one in the anterior-posterior direction and the other in the medial-lateral direction. Components of the GRF can be used to provide kinematic information about the activity in each of these directions. From Newton's second law, we know that the acceleration of an individual's centre of mass is directly proportional to the resultant force acting on their body. From this, acceleration, velocity and displacement can easily be obtained, as shown later in this chapter.

Given the importance of force and pressure measurement in the analysis of sport and exercise activities, technologies have been developed to accurately record these in many different environments (Hunter et al., 2005; Mornieux et al., 2006; Mills et al., 2010; Yu et al., 2016; Brauner et al., 2017). These technologies all incorporate some form of force transducer (also commonly known as force sensors). Whilst it is beyond the scope of this chapter to cover the physics and engineering principles behind all force sensing technologies, the chapter will provide an overview of force platforms, which are the most commonly used device both within a laboratory and in the field. Force platforms are normally located on the ground, to assess the foot – ground interface, but in certain scenarios they can also be located on a wall, on an elevated platform or within other sporting apparatus such as a diving platform (Lake and Lafortune, 1998; Galbraith et al., 2008; Ikeda et al., 2016). The chapter will also cover some of the commonly used pressure measuring devices which are used either to measure the pressure between the shoe and the ground (platform), or between the foot and the shoe (in-shoe).

From an historical perspective the first force-measuring device used in human motion is commonly attributed to Marey, who described a system in 1895 (see Nigg and Herzog, 2007). His apparatus employed pneumatic technology with air-filled tubes, which registered the pressure, and was thus related to the force exerted by the foot on the ground. This device was carried by the runners themselves and had many shortcomings. Subsequent investigators identified the advantage of having the measurement device independent of the performer and ideally being non-intrusive. This led to the development of force sensitive platforms, which were also designed to react to horizontal as well as vertical forces. One of the early examples of this approach is the 'trottoir dynamographique' (see Cavanagh, 1990) which used rubber bulbs that were compressed by the movement of the platform. The earliest known cousin of the current force platforms was developed and used by Elftman in 1938. This platform used springs to register the forces exerted by the runner against the ground (see Nigg and Herzog, 2007). One of the main limitations of this device was the magnitude of displacement necessary to compress the springs.

With advances in instrumentation and sensor technology, more sophisticated designs were developed which incorporate strain gauges and then later, piezo-electric crystals. These sensors allowed for rigid designs and much sturdier platforms. These platforms were non-deformable and allowed for high natural frequencies, which enabled the forces involved in high velocity impacts to be accurately recorded. One of the first commercially available force platforms was manufactured during the late 1960s by Kistler in Switzerland; this used piezo-electric crystal based technology. Advanced Mechanical Technology Incorporation (AMTI) later developed their own platform that used strain gauge transducers. This technology facilitated the measurement of a resultant force over a large contact area when compared to other technologies at that time. Both these platform types have gone through a number of modifications in recent years, with changes in sensor technology and computing power, and remain the most popular choice for use in sport and exercise applications.

Construction and operation of force platforms

Force platforms are normally rectangular in shape, are available in a variety of sizes and physical construction. Some models now have opaque tops (BP600900 GT or OR6-GT from AMTI) which allow viewing of the performer's movement from below.

A typical example of a commonly used force platform is the OR6–7-OP-1000 model from AMTI. This platform measures 464 × 508 × 82.55 mm and has a mass of 28.18 kg. It can measure forces up to 8896 N with an accuracy of ± 0.25 per cent of the applied load. This platform has a natural frequency (this is the resonate frequency of the force plate and is set by the mechanical properties of the plate itself) of up to 500 Hz (in the vertical direction) and a hysteresis (the difference in output for a given applied force during the loading and unloading) of ±0.2 per cent full-scale output. It uses strain gauge based technology with thin-walled cylindrical sensing elements. Each of these elements is instrumented with strain gauges, which are normally excited by a constant voltage supplied by the connected signal conditioner and amplifier. When a force is applied to the surface of the platform, strains occur in the walls of the supporting cylinders which, in turn, affects the electrical resistance of the strain gauges. This change in resistance and the output voltage are proportional to the forces being applied. Another popular platform is the Kistler 9281E, which has similar dimensions to the AMTI platform described above (600 × 400 × 100 mm) but a mass of only 16 kg. This model can measure forces up to 20,000 N in the vertical direction. The platform has a natural frequency of up to 1000 Hz and has ±0.2 per cent linearity (the maximum deviation of the force output curve from a specified straight line over the full measurement range). It has four built-in piezoelectric three component force sensors.

AMTI and Kistler force platforms have a top cover made of thick aluminum and are normally mounted on a solid metal base plate. They can provide information on six variables as shown in Figures 6.1 and 6.2.

In terms of an AMTI force plate (Figure 6.1), these six variables are the force components in the medial-lateral (*F*x), anterior-posterior (*F*y), and vertical (*F*z) directions along with the free moments *M*x, *M*y, *M*z about these axes. Regarding the Kistler force plate (Figure 6.2), the six variables are *F*x, *F*y and *F*z – the three components of the reaction force along the respective coordinate axes, *A*x and *A*z – the coordinates which identify the point of force application (commonly referred to as the Centre of Pressure (CoP) or centre of force), and *M*y – the free moment about the vertical axis.

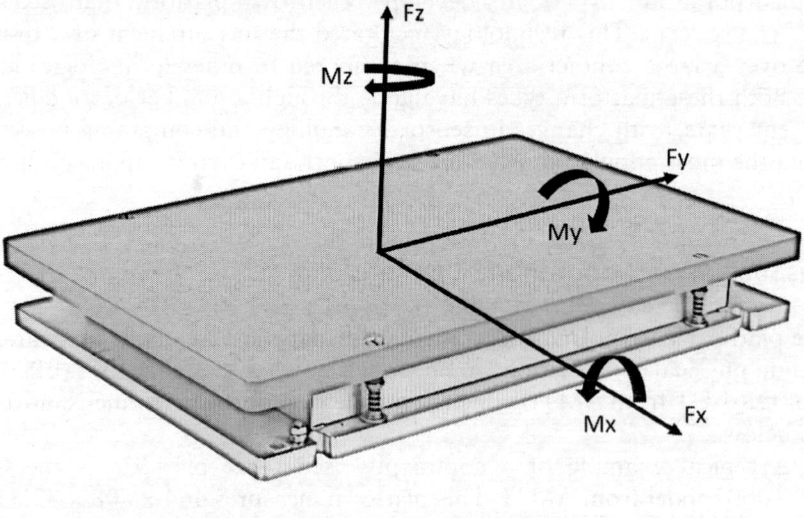

Figure 6.1 Schematic of AMTI force platform with superimposed axes.

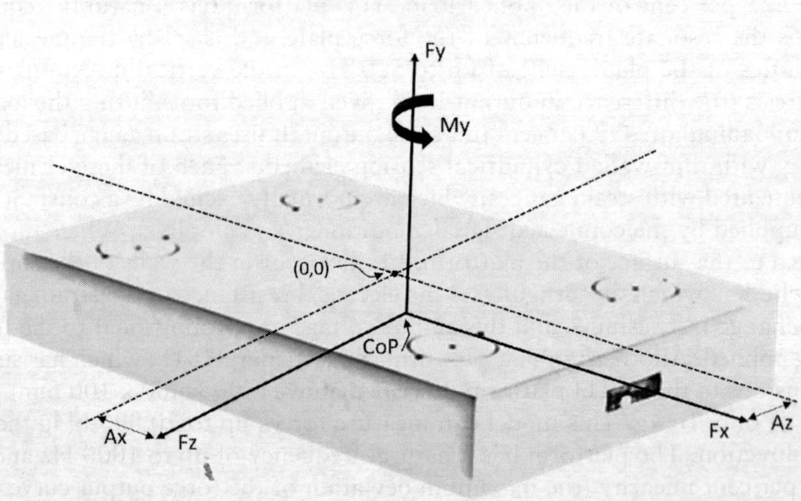

Figure 6.2 Schematic of Kistler force platform with superimposed axes.

Normally the Kistler force plate has four measuring elements which are situated in each of its four corners (Figure 6.2). Each sensing element is constructed in such a way that it is sensitive to forces along each of the X, Y or Z axes (the actual construction of the load cell depends on whether it uses the piezo-electric or strain gauge principle). The general principle of operation is that, when an external force F is applied, a reaction force is generated by the sensing elements to retain equilibrium (i.e. $\Sigma F = 0$).

The Centre of Pressure can provide information on the static and dynamic balance of an individual. The method for computing the CoP differs slightly between force plates. One can refer to the technical manual from the manufacturer for further details. However, it should be mentioned that in methodological considerations, and in experimental conditions, it is highly unlikely that one would need to manually calculate these values or export them for data reduction and synthesis. Nonetheless, there are some important issues that need to be addressed at this stage:

During any experimental set up, the axes have to identified and defined in an appropriate manner. Papers within the published literature vary in the way that they define these axes. Although the International Society of Biomechanics suggests that the vertical axis is termed Y, most force platform manufacturers, and many published papers, designate the vertical axis as Z.

The force plate measures the forces acting on it. It is important that we understand this and, in accordance with Newton's third law, we need to consider the reaction force which acts on the individual. To clarify, when a force is applied to the force platform by the individual, the sensing elements record this force which acts on the plate. However, the force acting on the individual is the 'reaction' force which is equal (in magnitude) and opposite (in direction) to the applied force. In essence, a downward force applied to the plate is recorded as an upward force acting on the body. The same principle applies to the horizontal forces and the free moment(s). Suitable corrections have to be made to the axis nomenclature (positive and negative directions can usually be specified within the system set up and analysis software).

The Centre of Pressure calculation makes an assumption that the external force is applied to the force platform in the same horizontal plane as that measured by the sensing elements. In many cases where force plate covers are used, such as indoor flooring or artificial running tracks, or when the geometry changes because of mounting considerations, corrections have to be applied. These corrections are often incorporated into the software used for system set up.

Technical specifications

A typical force plate system consists of the platform with sensing elements and an amplifier with a signal conditioning device. These are normally connected to a computer which controls the device and records the output.

The system should normally have a wide range of measurement, good measurement accuracy, good sensitivity and linearity, low hysteresis and a high natural frequency. In addition, the system should have high Centre of Pressure prediction accuracy and a low cross talk between various measurement channels (minimal interference of force applied in one direction with the measurement of force in an orthogonal direction). Normally one would also expect the platform to provide a good dynamic response during measurements, that is, it should be able to accommodate forces that change rapidly with time, such as impacts.

Whilst it is not easy to list typical values for force plate performance characteristics, given the range of commercially available platforms, the technical manuals all provide this basic information and one should exercise due diligence before choosing these plates, considering carefully their intended application. In terms of the range, typically one should look at measuring the GRF anywhere between 10 and 5000 N, with an accuracy of 0.25 per cent of the applied load. In addition, with advances in computing technology, the user has a variety of options to choose for data recording and reduction. Computing devices, normally a PC, need to have an appropriate operating system as specified by the manufacturer of the force plate.

Whilst there are some standalone systems, in most experimental setups the force plate is normally synchronised with other recording equipment. The synchronisation device or the option for synchronisation signals are inbuilt within the amplifier and the signal conditioning unit. This unit helps to convert the analogue signal to a digital data set. These are typically carried out by a 12- or 16-bit computing device or analogue-to-digital convertor (ADC). For a 12-bit device the signal is converted to 2^{12} (4096) data points. This is adequate for most force measurements as it indicates that a signal within the range of ± 5000 N can measured with a resolution of 2.4 N. This is better than the inherent noise within the signal, which can be seen if a base line measurement is amplified to cover a 10 N range. When the measured forces are relatively low, e.g. when no impact is involved, a lower range can be selected to improve the resolution of the force measurement. A 16-bit device provides considerably better resolution than a 12-bit device for a given force range (2^{16} – 65536 data points) and is therefore preferable where precise measurements of ground reaction force are required.

All commercially available force plates are supplied with their own proprietary software which provides various options for data reduction and presentation. However, the user should be aware that they will always have the option to export the raw data to other systems, such as motion capture systems, for reporting. Biomechanists will often choose this option as it provides them with further opportunities for analysing the data either in the time domain or the frequency domain.

Installation and calibration

Care should be taken when designing a laboratory and during the preparation for force plate installation. For typical sport and exercise testing, the platform should be mounted on a heavy rigid mass, typically a reinforced concrete base and isolated from the immediate building and other environmental vibrations.

Whilst most commercially available force platform systems are relatively easy to set up and use, the users should pay attention to the accuracy of the data produced. This has been questioned by previous research (Hall et al., 1996) and given that there are some old force plates still in working condition, one should check that they are regularly calibrated.

Commercially available systems are calibrated in the factory before they are shipped to the customer. However, users should be aware that the system setup at the user location includes the signal conditioning device and the analogue to digital converter. There might be occasions where these devices could introduce errors to the output. Calibration of the force plates has been a topic for debate and discussion within the scientific literature. To provide an overview, one should be aware that the accuracy of some of the force plate variables depends on local operating conditions. Moreover, the components of the GRF are dependent mainly on the sensing elements, but the other variables are more influenced by the operating conditions such as the mounting and the floor covering. The preferred and the easiest method for calibration *in situ* is to apply known weights. Various ways of doing this ways have been described by previous researchers (Bobbert and Schamhardt, 1990; Hall et al., 1996; Flemming and Hall, 1997; Middleton et al., 1999). This is an easy approach if the whole system needs to be checked for its measurement accuracy along the vertical axis. However, the calibration of the horizontal axes is more complicated. Within the methodology described by Hall et al. (1996), the calibration force was produced by a suspended weight applied in a horizontal direction by a system of 'frictionless' pulleys. The researchers observed that, even with a factory calibrated force plate with a small cross talk, a large vertically applied force can negatively influence the smaller horizontal forces (anterior-posterior and/or medial-lateral) in such a way that they show higher than normal forces. Another area which has been of interest to previous researchers is the accuracy of the location of the CoP (Bobbert and Schamhardt, 1990; Middleton et al., 1999; Chockalingam et al., 2002). Chockalingam et al. (2002) indicated that a threshold of 113 N was required in order to obtain a CoP coordinate value within a distance of 0.3 mm of its mean value. However, the accuracy of the location deteriorated toward the edges of the platform.

A dynamic calibration of a force platform is far more challenging than a static one. Fairburn et al. (2000) proposed a procedure for conducting a dynamic calibration based on a 20 kg oscillating pendulum of known inertial properties. This approach enabled them to compare the theoretical forces to the measured ones. Although there was no significant difference between the predicted and actual vertical forces, differences were found for the horizontal components.

Later, Cedraro et al. (2008) reported an algorithm which estimated the local or global six-by-six re-calibration matrix of any force plate through a least-squares optimisation method. This algorithm was tested and validated for several types of force plates in their subsequent paper (Cedraro et al., 2009). This approach provided a much better estimation of the CoP after recalibration. Current commercially available systems benefit from improved calibration procedures due to some of the published work in this area (ASTM F3109–3116).

General operation

Commercial force plates are delivered with software which not only helps with setup, calibration and data capture but also basic data analysis and presentation. This software also helps in choosing the sampling rate. Whilst the detailed concepts behind sampling rate can be found in Chapter 9 and elsewhere (e.g. Antonsson and Mann, 1985, Bartlett, 2014), the main decision should be based on the activity that is being recorded. Many of the synchronised systems for gait analysis have a default of 100 Hz. However, if one looks at the literature relating to sports activities, sampling rates ranging from 500–2000 Hz are often reported (Bartlett, 2014).

As indicated earlier, the force plate can provide information on the three components of the GRF, the moments around these axes and the free moment. These values are normally provided in the time domain and their patterns differ according to the type of activity being recorded. Figures 6.3 and 6.4 illustrate ground reaction force data for the stance phase of walking and running, respectively. The force values have been normalised to bodyweight (Force (N)/ Bodyweight (N)).

The vertical ground reaction force is the largest of the three ground reaction force components. For walking (Figure 6.3(c)) the first peak in the vertical ground reaction force relates to the loading of the foot during the initial contact phase of stance, the subsequent trough reflects movement of the body over the stance limb, and the second peak relates to the push off phase of gait. The peak vertical forces during walking, while dependent on walking speed (Schwartz et al., 2008), are normally ~ 1.2–1.5 times bodyweight. The pattern of the vertical ground reaction force for running (Figure 6.4(c)) is substantially different to that of walking. By definition, in running there is a flight phase, where both feet are off the ground, compared to the single (one foot in contact with the ground) and double support (both feet in contact with the ground) phases in walking. As the acceleration of the centre of mass is greater in running than walking the vertical forces are increased, with typical peak forces of 2–3 times bodyweight (Nilsson, 1989).

The anterior-posterior ground reaction force (Fy) data (Figures 6.3(a) and 6.4(a)) show an initial force acting posteriorly. This force is interpreted as the 'braking force' acting from initial foot contact through to mid stance. We then see an anterior or 'propulsive force' which acts from mid-stance through to toe off. It is important to record the direction of the movement in relation to the axes of the force plate, as this affects how the data are displayed. If the person whose data are presented in Figures 6.3 and 6.4 had walked across the force plate in the opposite direction, then the Fy data would be inverted, that is, the posterior (braking) force would be positive and the anterior (propulsive) force negative. The medial-lateral force (Fx) (Figures 6.3(b) and 6.4(b)) is the smallest force, as acceleration of the person's mass centre in this direction is much lower than in the other two. Initially, there is a lateral force as the foot is loaded after which the forces moved medially as the body moves over the limb.

The vertical ground reaction forces experienced during landing from a drop or a jump generally exceed those experienced during walking or running.

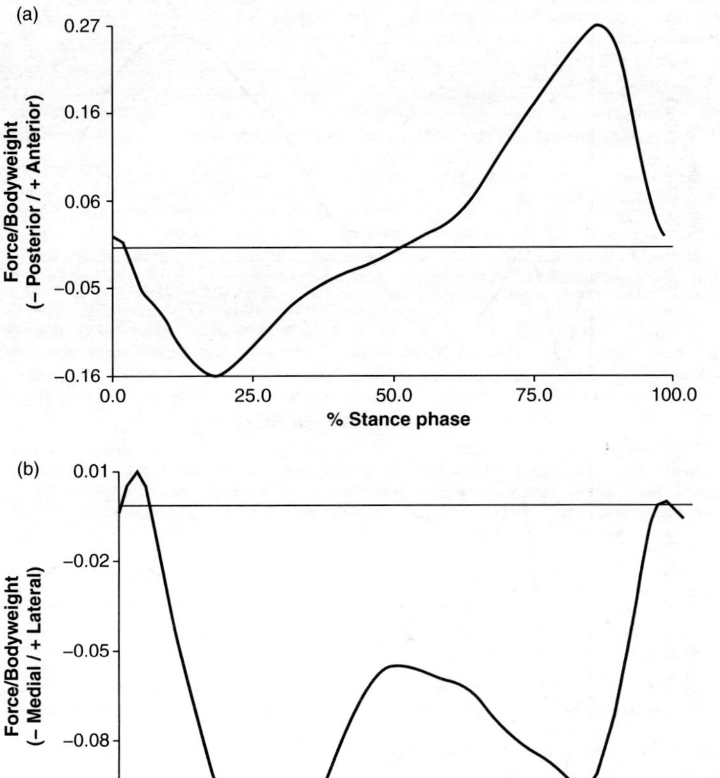

Figure 6.3 Sample force data for stance phase of walking: (a) *Fy*, (b) *Fx* and (c) *Fz*.

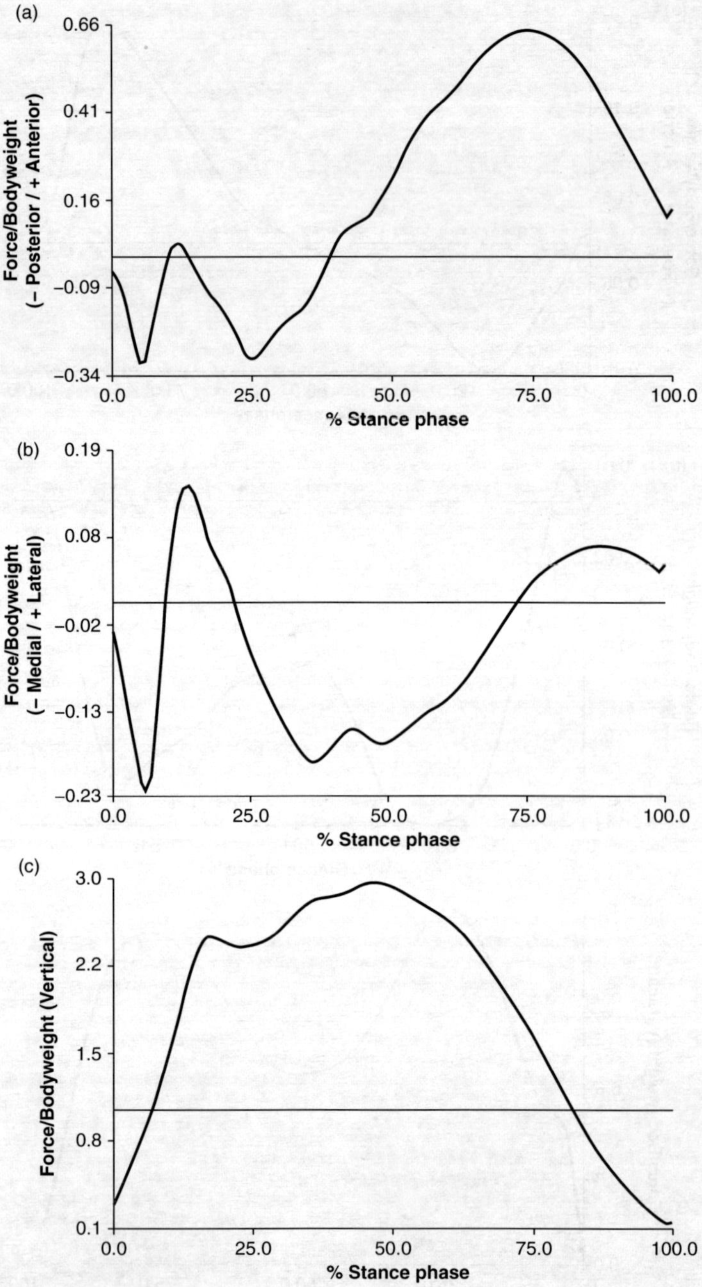

Figure 6.4 Sample force data for stance phase of running: (a) *Fy*, (b) *Fx* and (c) *Fz*.

Figure 6.5 provides sample data for a single leg jump landing from a 30 cm high box. Note that the peak vertical force is greater than five times bodyweight. During sport specific movements such as cutting (side-stepping) manoeuvres, the direction of travel is not normally aligned with an axis of the force plate and hence one may wish to calculate the resultant horizontal force, that is, the

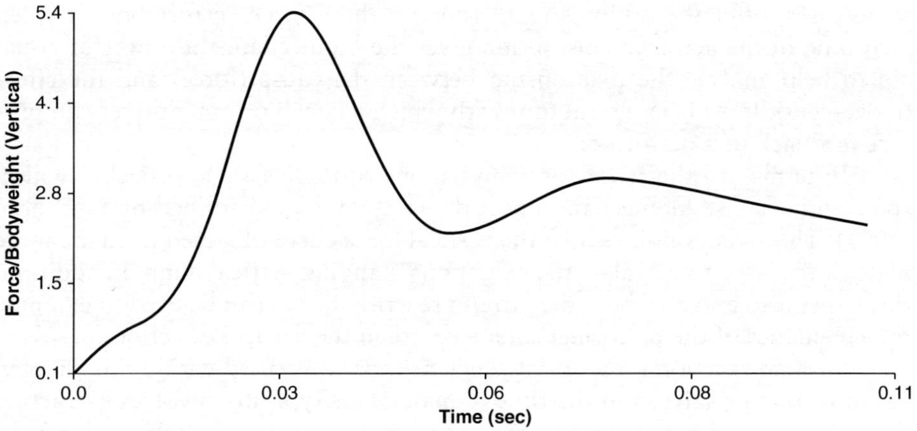

Figure 6.5 Sample vertical ground reaction force data for a single leg jump landing from a 30 cm high box.

total frictional force. This is simply done by computing the vector sum of the horizontal forces, Fx and Fy as follows:

$$Resultant\ horizontal\ (friction)\ force\ = \sqrt{\left(Fx^2 + Fy^2\right)} \tag{6.1}$$

Previous papers have indicated that the Centre of Pressure coordinates (Ax and Az in Figure 6.2) are accurate only when the vertical force exceeds a specified level (Chockalingam, 2002). However, improvements in the technology, the factory calibration procedures and the introduction of standards (ASTM F3109) have all contributed to an improved accuracy of some commercially available force plates.

Another important consideration is the placement of the force plate(s). Most biomechanics/gait laboratories will have two or more plates installed. Whilst an infinite number of placement options exist, normally, for gait analysis, one would place the force plates adjacent to each other, in the direction of travel with a slight offset. If three plates are available, it is useful to locate two of them side-by-side with no offset. This set-up is particularly useful when collecting data during activities such as jump landing, golf swing and pistol shooting. Some force plate manufacturers now offer a system where one can move the force plates with a set pattern. This option might prove useful for a laboratory set up where several researchers collaborate on a variety of projects.

Analysis of data

With the current advances in the computer technology and software capabilities, the data reduction, analysis and detailed interpretation of force data is fairly straightforward. Most traditional interpretation is carried out in the time domain using graphs similar to those shown in Figures 6.3 to 6.5. However, most software packages also allow the visualisation of the force vector and, in some cases, offer the option to superimpose these force vectors onto a video recording of the activity. This option gives the biomechanist a powerful visual aid to help analyse the relationship between the cause (force) and the effect (movement). It is also a useful tool with which a coach or practitioner can provide feedback to a performer.

In terms of other types of analysis, one of the classic approaches within sport and exercise biomechanics uses the methodology described by Linthorne (2001). This paper outlines how the vertical force curve obtained from the force plate can be used to calculate the height of a standing vertical jump. In addition, this paper also explains how the ground reaction forces can be used to examine the kinematics of the performer's mass centre in the vertical direction.

Ground reaction force curves, such as those presented within this chapter, are most often analysed in the time domain. This typically involves extracting discrete force and time values from the curves, such as the peak vertical ground reaction force and duration of the braking phase, and calculating variables as a function of time, for example loading rate (described later in this section). An alternative method of analysis is frequency domain analysis. This refers to the inspection of the force signal with respect to its frequency content, rather than looking at it as a function of time. Fourier analysis is the method used to convert the force data from their original time domain to the frequency domain. This type of analysis, although not as popular as time domain analysis, may reveal important findings that were not apparent in the time-series data. Shorten and Mientjes (2011) utilised frequency domain analysis to examine ground reaction forces during running. They examined the frequencies present in the initial transient peak of the vertical ground reaction force (Fz_1) during running while wearing shoes with different amounts of cushioning. Through the use of frequency domain analysis this study demonstrated that the use of Fz_1 as a measure of heel impact was not valid. They found that Fz_1 is a composite that includes high frequency (heel impact) loads but that it also contains low frequency force components originating both in the heel and distal forefoot.

In addition to the GRF components there are other variables which prove to be useful in describing and understanding physical activity. Two force-related variables often measured in sport and exercise activities are the impulse and the loading rate.

Impulse

The impulse is defined as the area under the force–time curve. This measure is of great importance in sport and exercise activities as it literally represents the change in momentum experienced by the person who generated the force curve. This association is referred to as the impulse-momentum relationship (see for example, Bartlett, 2014, for more detail on this key concept). Impulse can be calculated easily in a simple spread sheet, e.g. MS Excel, or by writing simple software codes. The two common numerical methods for calculating the integral (the area under the force curve) are Simpson's rule and the Trapezium rule. Although the former is usually more accurate, the latter is easier to calculate and will provide a good approximation of the impulse provided that the time interval between successive force data samples (1/sample frequency) is sufficiently short. Over a number of samples N, the area under the force – time curve is given as follows:

$$\text{Area} = \sum_{N}^{i=1} (F_i + F_{i+1}) \frac{\Delta t}{2} \qquad (6.2)$$

Where *F*i = the value for force at period i, and *Δ*t = the time interval between force samples.

Impulse has been used effectively in the description of the anterior-posterior force component during running (Hunter et al., 2005). In Figure 6.4(a), the initial negative areas represent the braking impulse. The impulse-momentum relationship tells us that the runner will lose horizontal momentum (and therefore speed) during this phase. The subsequent positive area represents the propulsive impulse. The runner will therefore regain some momentum during this phase of the ground contact. At a constant running speed, the braking and propulsive impulses should be very similar in magnitude, indicating that the runner would leave the force platform (toe-off) with similar forward momentum to when they contacted it (heel-strike). If the propulsive impulse is greater than the braking impulse, the runner would gain horizontal speed during the ground contact. Conversely, if the braking impulse is greater than the propulsive impulse, the runner would lose horizontal speed. Visual inspection of the anterior-posterior force curve can be used to identify whether a participant is changing their pattern of movement, e.g. over-striding, in order to hit the force platform.

Loading rate

Many publications have reported the loading rate or the rate of change of force associated with running and jumping related activities. It has been argued that this variable represents the shock imposed on the musculoskeletal system during the impact associated with heel-strike in running (Hennig et al., 1993; Laughton et al., 2003). There are various ways loading rate may be calculated, including finding the average slope between two arbitrary levels of the peak force (e.g. 20 to 90 per cent), and computing the instantaneous value through numerical differentiation of the force curve. Woodward et al. (1999) investigated the effects of several computational methods used for the determination of loading rate. They found that the method most sensitive to differences between two clinical groups was the differentiation method based on the central difference formula (equation 6.3).

$$\text{Loading rate at point } i = (F_{i+M} - F_{i-M})/(2M\,\Delta t) \qquad (6.3)$$

Where, *M* is the spread of data points either side of the central point (*i*). Woodward et al. (1999) found that the loading rate reduced as *M* increased but values of *M* = 1–19 could be used.

As shown by Newton's second law of motion, the force recorded by a force plate is a direct representation of the performer's centre of mass acceleration. It is therefore easy to compute the acceleration and, subsequently, changes in the velocity and displacement of the performer's centre of mass, directly from force platform data. This can be illustrated with a simple example. Consider a person with a body mass, *m*, standing on a force platform in the process of performing a countermovement jump (Figure 6.6).

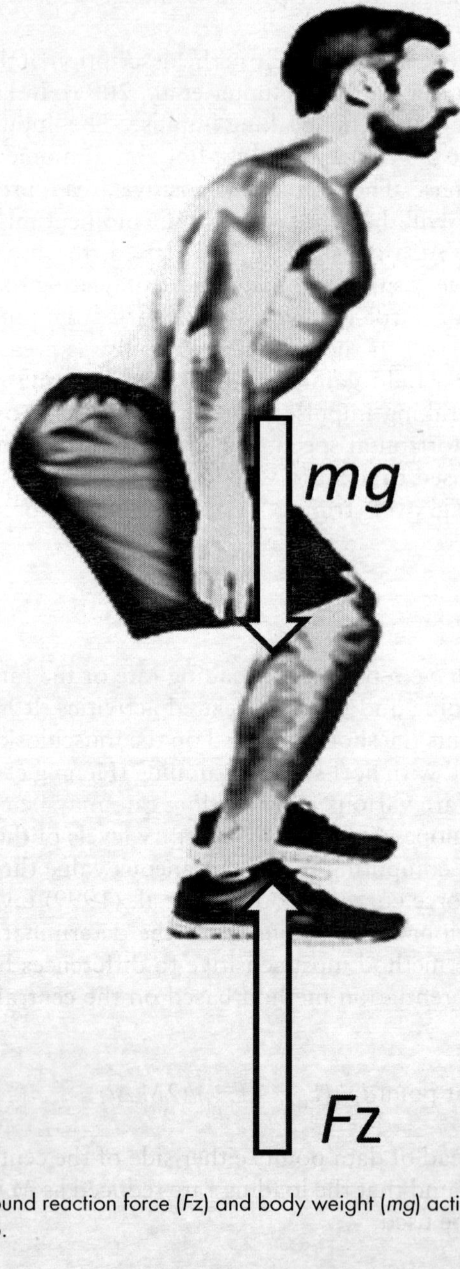

Figure 6.6 Vertical ground reaction force (*Fz*) and body weight (*mg*) acting on a performer during a standing vertical jump.

If we consider forces acting in only the vertical direction, we have the person's weight due to gravity (mg) acting downward and the vertical ground reaction force (F_Z) acting upward. The net vertical force acting on the person (ΣF_{VERT}) is then given by:

$$\Sigma F_{VERT} = F_Z - mg \tag{6.4}$$

But, as $\Sigma F_{VERT} = m\,a$ (from Newton's second law)

Then $$a = \Sigma F_{VERT}/m \tag{6.5}$$

Thus the vertical acceleration of the person's mass centre (a) can be obtained at any instant during the contact with the force plate. As acceleration is related to the change in velocity from an initial (v_i) to a final value (v_{i+1}) as follows:

$$v_{i+1} = v_i + \bar{a}\,(t_{i+1} - t_i) \tag{6.6}$$

where, ā is the mean vertical acceleration between i and i+1.

Then, velocity is related to the change in displacement from an initial (d_i) to a final (d_{i+1}) position as follows:

$$d_{i+1} = d_i + \bar{v}t \text{ (where } \bar{v} \text{ is the mean vertical velocity between } i \text{ and } i+1) \tag{6.7}$$

Equations 6.5, 6.6 and 6.7 enable us to compute the acceleration, velocity and displacement of the individual, respectively, from the net force acting. A detailed description of the process of calculating velocity from acceleration, and displacement from velocity, termed integration, are described elsewhere (e.g. Bartlett, 2014).

Interpretation of a force curve

A typical vertical ground reaction force-time curve for a countermovement jump is presented in Figure 6.7. Also shown are the associated acceleration, velocity and displacement curves obtained from the force curve using integration. The curves presented are from the initiation of the countermovement jump until the feet leave the ground (take-off) for the flight phase of the jump. Initially the vertical ground reaction force drops below bodyweight (~ 800 N), this phase is referred to as the unloading phase. Force decreases below bodyweight in this phase as the centre of mass moves in the same direction (moving downwards) as the acceleration due to gravity. Acceleration and velocity are increasing in a negative direction. Note that the three time points where acceleration is zero coincide with when the net force is zero ($Fz = mg$), and these correspond with the instances of maximum, minimum or zero velocity. The unloading phase is followed by the braking phase. In this phase the performer's centre of mass continues to move downward, as indicated by the negative velocity, but slows

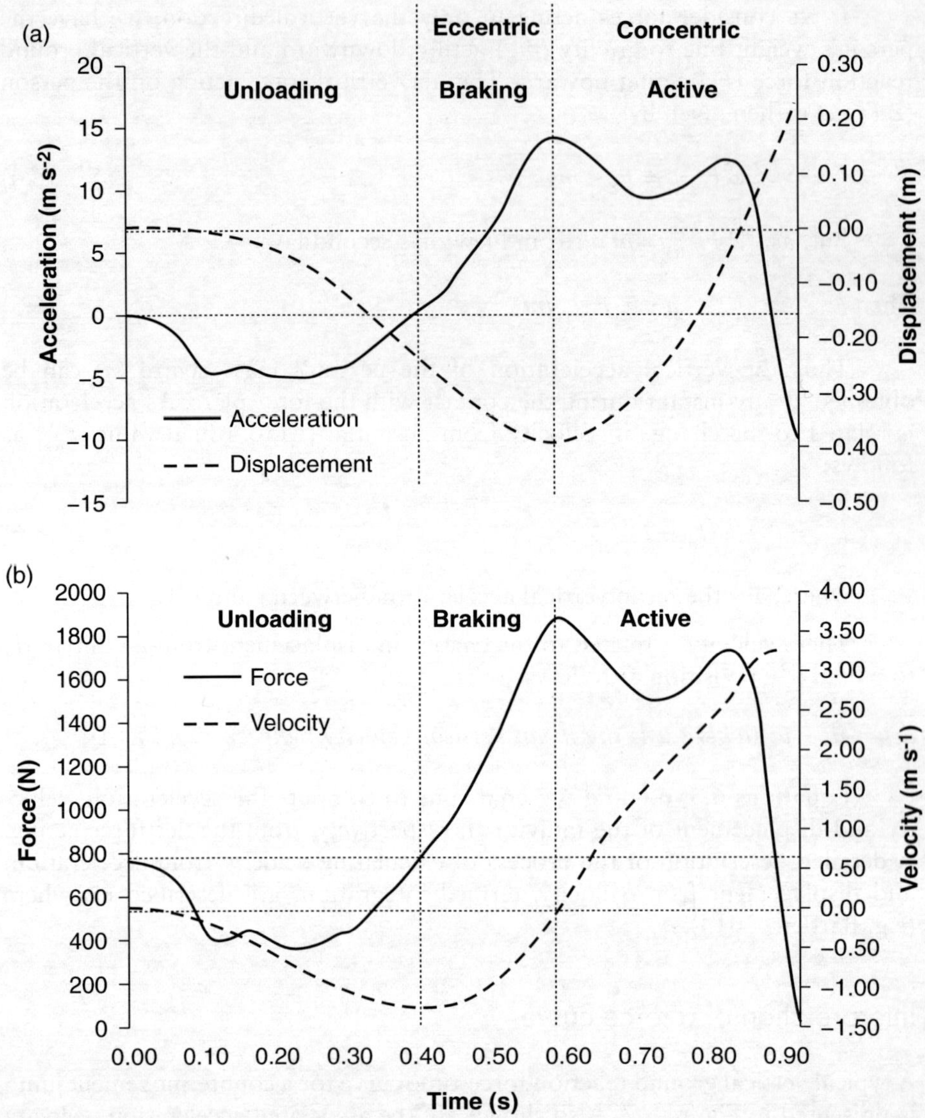

Figure 6.7 Sample derived acceleration and displacement data (a) and vertical ground reaction force and derived velocity data (b) for a countermovement jump.

down, and eventually stops, due to the ground reaction force rising above their bodyweight (>800 N). The braking phase ends at the bottom of the countermovement squat. Here, the centre of mass is at its minimum displacement and has zero velocity. The final phase is the active phase, which commences with the initiation of the upward movement of the centre of mass and we thus see a change from negative to positive velocity. For most of this phase, the ground reaction force exceeds bodyweight, acceleration is therefore positive and upward velocity increases. The performer's centre of mass moves back toward, and then past, its start position. At the end of this phase, just prior to take off, the ground reaction force drops below bodyweight causing the acceleration to become negative and the upward velocity to start to decrease.

Applications of force platforms

Force plates have been used within sport and exercise biomechanics for a variety of reasons and in numerous ways. The variables measured from forces plates have become a standard and reliable outcome measure for various studies.

For example McNair et al. (2000) used force plate data to demonstrate that giving performers verbal instructions relating to the kinematics of their knee and ankle joints is an effective method of lowering the vertical ground reaction forces associated with landing from a jump. McLean (1994) demonstrated that data generated from a force plate can effectively be used in the assessment of a hurdling technique. The author highlighted the potential of ground reaction force data for diagnosing technical shortcomings and how such data could be used for effective intervention by a coach. In addition, because of the nature of force plates, the data collection and analysis can be performed in conditions similar to the training environment, and immediate feedback is possible. This makes the force plate a popular choice for sports scientists.

With a view to providing a full description of the ground reaction force characteristics of professional basketball players, McClay et al. (1994) studied the forces experienced by the players during their movements. The results highlighted that certain common basketball movements, such as jump landings and shuffling, resulted in absolute and relative forces considerably higher than those previously reported. Studies such as this not only describe sport specific movements, but help in the development of new ideas and hypotheses that extend the knowledge within the sport; aiding injury prevention, rehabilitation and coaching practice.

Force plate data have also been used as a main outcome measure for exercise intervention based clinical studies. For example, Kelley et al. (2012) indicated that low bone mineral density and subsequent fractures are a major public health issue in postmenopausal women. They conducted a meta-analysis to examine the effects of exercises such as walking and strength training on bone mineral density in the femoral neck and lumbar spine of postmenopausal women. Articles within this meta-analysis used force plates not only to assess ground reaction forces, but also to provide quantified and controlled exercise protocols.

As outlined earlier, data from force plates can be used simply to quantify external forces, processed to obtain derived variables, for example loading rate, and integrated to obtain centre of mass kinematic quantities. In addition, the data are also widely used to compute joint forces and moments (see Chapter 11), which normally requires the integration of the force plate with an appropriate motion analysis system (see Chapter 5).

PRESSURE MEASUREMENT

Pressure measurement systems provide information on the pressures between the contact points of the body and the supporting surfaces, for example the ground, during activities. Pressure is defined as the ratio of force to the area over which that force is being applied (Equation 6.8). This force is normally considered to be acting in a direction perpendicular to that of the surface to which it is it is applied.

$$Pressure = \frac{Force}{Area} \qquad (6.8)$$

When pressure measurement is related to the foot it is termed plantar pressure, as measurement is usually between the plantar surface of the foot and the ground. Plantar pressure measurement has a wide range of applications in sport and exercise biomechanics. For example, assessing the effects of different insoles in football boots (Nunns et al., 2016) and running shoes (Hennig and Milani, 1995), examining the relationship between weight transfer and club head speed in golf (Pataky and Lafortune, 2015) and assessing the effect of different running surfaces (Tessutti et al., 2012). Figure 6.8 shows example plantar pressure profiles for two runners, one with a rearfoot and the other with a forefoot initial contact running pattern.

Historically, a range of techniques have been used to quantify pressures, including pressure sensitive film, pedobaraographs (which utilise an elastic mat, glass plate, mirror and camera to measure pressure) and electronic sensors. Electronic sensors are the predominant approach used now; when compressed, these sensors measure the applied force and, using the known area of the sensor, convert the measured force to pressure. Detailed information on the various types of sensors is available (see for example, Nigg, 2006; Rosenbaum and Becker, 1997).

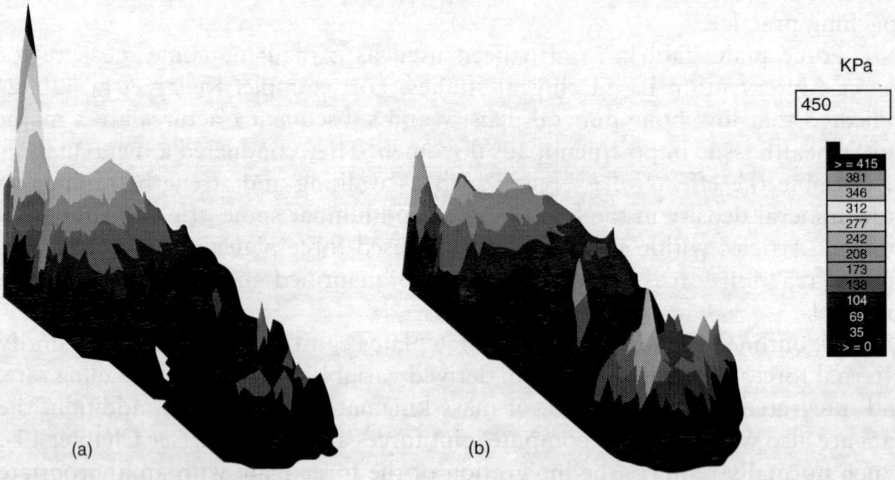

Figure 6.8 In-shoe plantar pressure measurements of two individuals running at 2.8 m s^{-1} on a treadmill: (a) initial foot contact on the forefoot and (b) initial foot contact on the rearfoot.

Technical specification

While considerable advances have been made since the first available pressure measurement systems, current systems still have limitations and can be affected by the heat and humidity of the testing environment (Cavanagh et al., 1992). It is important to understand the technical specifications of the pressure measurement system you are using to ensure it is capable of providing the required information. The system should exhibit low hysteresis, which relates to the maximum discrepancy in pressure readings when a comparison is made between loading and unloading of the sensors. The relationship between the force applied to the sensor and the output signal should ideally be linear. The system should also exhibit low cross talk between sensors; cross talk is the undesired activation of unloaded sensors when pressure is applied to neighbouring sensors. When considering the high speeds associated with many activities in sport and exercise, the sampling frequency is a key consideration. When recording walking, a sampling frequency of 50–100 Hz is acceptable; for higher speed activities, such as running, a minimum sampling frequency of 200 Hz is recommended. The pressure range defines the upper limit of the pressure the system is capable of measuring; the system must have the capacity to record the peak pressures created by the activity of interest, without being overloaded. Calibration of pressure measurement systems is essential to ensure accurate measurement. Some systems are calibrated by the manufacturer whilst others require the user to perform a calibration prior to each testing session.

Types of pressure sensors

There are a wide range of sensor arrangements now available to allow pressure measurement between different interfaces. These include an extensive number of systems which enable the measurement of the interaction between the foot and the ground (i.e. pressure mats, in-shoe systems and instrumented treadmills). Systems are also available to assess hand grip pressure, sitting and lying pressures in a chair/wheelchair/bed, and sensor arrangements to assess the fit of prostheses. Small individual sensors and customised sensor arrangements for specific tasks are also available. Knowledge of the spatial resolution (the size of the sensors) of the pressure system is very important. The sensors need to be small enough to measure pressures over the area of interest, for example the small anatomical structures of the foot such as the metatarsal heads. Sensor size is particularly important when studying pressure under small feet, such is in children. Currently available pressure platforms systems typically have 1–4 sensors per cm^2. For in-shoe pressure systems the number of sensors per insole is dependent on the shoe size and can range from as low as 24 to over 1000 sensors per insole.

Pressure mats or platforms are predominately used for the assessment of barefoot plantar pressures during static and dynamic activities. These systems are available in a range of sizes, with the smallest allowing for the capture of a single foot strike during walking and the larger systems capable of recording multiple gait cycles. In-shoe pressure systems allow for the measurement of the pressures between the foot and the shoe, or the foot and a foot orthosis. A challenge when using these systems is fitting the sensors into the shoe where there is limited space. The sensors need to be as thin as possible; they usually range between 0.15 mm and 3.5 mm in thickness. In-shoe systems are linked to a computer where the data are recorded via either a wired or wireless connection. For wired systems, the number of steps that can be recorded in one trial is limited by the length of the cables. While this is not an issue for the wireless systems, many of these require the participant to wear a belt around their waist containing a data logger and battery, which may be restrictive depending on the activity being performed. Instrumented treadmills capable of measuring pressures are now available and are particularly useful if the laboratory has limited space to record plantar pressures during walking or running.

Practical considerations for data capture

Pressure measurement is generally conducted in a laboratory environment. A number of factors need to be considered to ensure that the participant performs the activity of interest as normally as possible and that valid and reliable pressure data are obtained:

- Participants should not 'target' the pressure platform, which means they should not shorten or lengthen their walking/running stride in order to make contact with the platform. Techniques used to reduce targeting include positioning mats (with the same thickness as the height of the pressure platform) before and after the platform (Naemi et al., 2012) or embedding the platform in the floor. Targeting can also be reduced by asking the participant to walk across the platform with their head up and looking forward, so they are not aware of when they are walking on the platform.
- The number of steps taken before the participant makes contact with the pressure platform should be standardised, as previous research has shown that the number of steps taken before contacting the platform can affect the pressure values recorded (Bus and de Lange, 2005). Figure 6.9 provides examples of step protocols for data capture where the participant contacts the platform with their first or second step.
- Walking/running speed should be monitored when measuring plantar pressures as an increase in speed generally results in an increase in pressures under the foot (Rosenbaum et al., 2013). Should the testing protocol involve multiple testing sessions the same walking/running speed should be maintained across sessions if pressure data are to be compared across these sessions. Also, the participant's footwear and socks worn should be standardised across sessions.
- In-shoe pressure sensors are susceptible to creasing or bending, as they are fitted within the limited space within footwear, which can result in measurement artefacts. A crease or bending of the insole within the shoe can

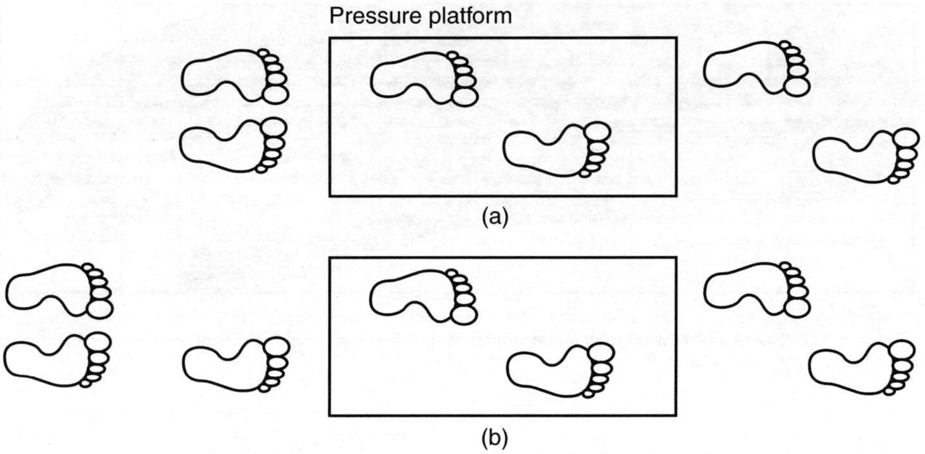

Figure 6.9 Data capture setup for a one-step (a) and two-step (b) protocol.

result in a sensor outputting a very high pressure value which is unrelated to the plantar pressures. It is essential to ensure correct fitting of the sensor within the footwear prior to testing and to monitor the outputs during data capture to identify potential creases/bending of the sensor.

- In-shoe pressure sensors are more susceptible to degradation than platform systems as these sensors can be subjected to bending within the shoe. The heat and humidity generated within the footwear can also damage the sensors. Sensors degrade quicker when subjected to higher forces, so their life-span will be shorter if they are used for measuring relatively high force activities such as running and jumping, compared to walking.
- Chevalier et al. (2010) recommended that pressure values should not be compared between different pressure measurement systems, and between platform and in-shoe systems. Pressure systems measure the force that is perpendicular to the sensor surface; for a pressure platform this can be considered the vertical force. However, in in-shoe systems the force can only be considered as the vertical force for the portion of the stance phase when the foot is parallel to the ground.

Pressure measurement systems offer a range of output measurements including average and peak (maximum) pressure, pressure time integral (area under the average pressure-time curve), force, impulse (area under the force-time curve), contact area and centre of pressure. Pressure measurement system software allows users to examine pressures in more detail in specific regions, termed regions-of-interest (Figure 6.10). Within the software the user can apply boxes/masks to specific regions. This approach is frequently used to examine the effect of different footwear or foot orthoses on specific regions on the foot.

In conclusion, pressure measurement systems allow valuable kinetic information to be recorded for both foot – ground interaction and other loads acting on the body. Some of the key issues in selecting a pressure measurement system for a specific application have been discussed, including the importance of

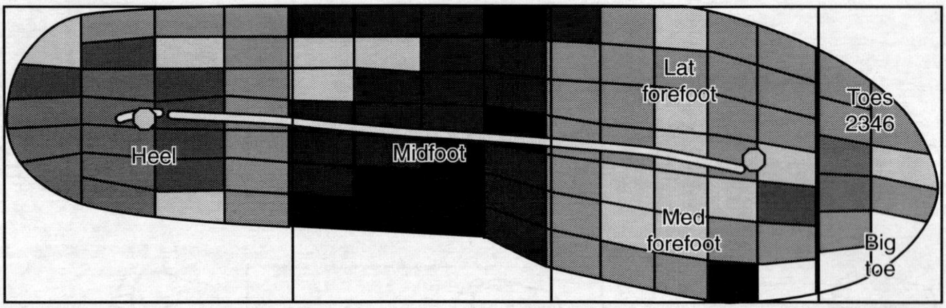

Figure 6.10 Six regions of interest: heel, midfoot, left forefoot, right forefoot, big toe, toes (Novel GmbH, Munich, Germany).

matching the sensing arrays to the characteristics of the loads expected. Where possible, the frequency content and pressure range associated with the activity of interest should be identified before deciding which sensor type and sampling frequency to use.

REPORTING PROCEDURES

As good practice, any report or a publication should include enough information for a third party to recreate the experiment and achieve the same results as the ones reported within the paper. In terms of force and pressure data, the following information should always be included when reporting a study (Lake and Lees, 2008):

Force platform

- Manufacturer and model of the system (including dimensions of the plate and the type of amplifier).
- Manufacturer, model and resolution of the analogue-to-digital convertor or other device used to sample the data.
- Calibration method (any user modification of manufacturer calibration curves).
- Range settings.
- Sampling frequency.

Pressure measurement system

- Manufacturer and model of the system.
- Spatial resolution (number of sensors per cm^2).
- Sampling frequency.

- Maximum loading range (with any sensor material or design modifications).
- Number of sensors or size of array.
- Thickness of the sensors.
- Resistance to bending artefact (given as a radius of curvature).
- Description of the interface where pressure is being measured.

Processing and analysing force and pressure data

- Spatial alignment of the force platform reference system to global coordinate system.
- Method used to synchronise the force platform or pressure system with other data acquisition systems, e.g. motion analysis system.
- Manufacturer and version of software used for processing data.
- Details of any masks used to define regions for pressure analysis.
- Method used to smooth/filter the data.
- Method used to obtain any derivative or integral data (e.g. loading rate, impulse).
- Definitions of the dependent variables (parameters) being quantified, including their SI units.

REFERENCES

Antonsson, E. K. and Mann, R. W. (1985) 'The frequency content of gait', *Journal of Biomechanics*, 18: 39–47.

ASTM F3109–3116 (2016) 'Standard test method for verification of multi-axis force measuring platforms', ASTM International, West Conshohocken, PA, www.astm.org. DOI: 10.1520/F3109–3116

Bartlett, R. M. (2014) *Introduction to Sports Biomechanics: Analysing Human Movement Patterns*, 3rd edn. London: Routledge.

Bobbert, M. F. and Schamhardt, H. C. (1990) 'Accuracy of determining the point of force application with piezoelectric force plates', *Journal of Biomechanics*, 23: 705–710.

Brauner, T., Pourcelot, P., Crevier-Denoix, N., Horstmann, T. and Wearing, S. C. (2017) 'Achilles tendon load is progressively increased with reductions in walking speed', *Medicine and Science in Sports and Exercise*, 49(10): 2001–2008. doi: 10.1249/MSS.0000000000001322.

Bus, S. A. and de Lange, A. (2005) 'A comparison of the 1-step, 2-step, and 3-step protocols for obtaining barefoot plantar pressure data in the diabetic neuropathic foot', *Clinical Biomechanics*, 20: 892–899.

Cavanagh, P. R. (ed.) (1990) *Biomechanics of Distance Running*, Champaign, IL: Human Kinetics.

Cavanagh, P. R., Hewitt, F. G. and Perry, J. E. (1992) 'In-shoe plantar pressure measurement: a review', *The Foot*, 2: 185–194.

Cedraro, A, Cappello, A. and Chiari, L. (2008) 'A portable system for in-situ re-calibration of force platforms: theoretical validation', *Gait and Posture*, 28: 488–494.

Cedraro, A, Cappello, A. and Chiari, L. (2009) 'A portable system for in-situ re-calibration of force platforms: experimental validation', *Gait and Posture*, 29: 449–453.

Chevalier, T. L., Hodgins, H. and Chockalingam, N. (2010) 'Plantar pressure measurements using an in-shoe system and a pressure platform: a comparison', *Gait and Posture*, 31: 397–399.

Chockalingam, N., Giakas, G. and Iossifidou, A. (2002) 'Do strain gauge platforms need in situ correction?', *Gait and Posture*, 16: 233–237.

Fairburn, P. S., Palmer, R., Whybrow, J., Felden, S. and Jones, S. (2000) 'A prototype system for testing force platform dynamic performance', *Gait and Posture*, 12: 25–33.

Flemming, H. E. and Hall, M. G. (1997) Quality framework for force plate testing, *Journal of Engineering in Medicine*, 211: 213–219.

Galbraith, H., Scurr, J., Hencken, C., Wood, L. and Graham-Smith, P. (2008) 'Biomechanical comparison of the track start and the modified one-handed track start in competitive swimming: an intervention study', *Journal of Applied Biomechanics*, 24:307–315.

Hall, M. G., Flemming, H. E., Dolan, M. J., Millbank, S. F. D. and Paul, J. P. (1996) 'Technical note on static in situ calibration of force plates', *Journal of Biomechanics*, 29: 659–665.

Hennig, E. M. and Milani, T. L. (1995) 'In-shoe pressure distribution for running in various types of footwear', *Journal of Applied Biomechanics*, 11: 299–310.

Hennig, E. W., Milani, T. L. and Lafortune, M. (1993) 'Use of ground reaction force parameters in predicting peak tibial accelerations in running', *Journal of Applied Biomechanics*, 9: 306–314.

Hunter, J. P., Marshall, R. N. and McNair, P. J. (2005) 'Relationships between ground reaction force impulse and kinematics of sprint-running acceleration', *Journal of Applied Biomechanics*, 21: 31–43.

Ikeda, Y., Ichikawa, H., Nara, R., Baba, Y., Shimoyama, Y. and Kubo, Y. (2016) 'Functional role of the front and back legs during a track start with special reference to an inverted pendulum model in college swimmers', *Journal of Applied Biomechanics*, 32: 462–468.

Kelley, G. A., Kelley, K. S. and Kohrt, W. M. (2012) 'Effects of ground and joint reaction force exercise on lumbar spine and femoral neck bone mineral density in postmenopausal women: a meta-analysis of randomized controlled trials', *BMC Musculoskeletal Disorders*, 13: 177.

Lake, M. J. and Lafortune, M. A. (1998) 'Mechanical inputs related to perception of lower extremity impact loading severity', *Medicine and Science in Sports and Exercise*, 30: 136–143.

Lake, M. J. and Lees, A. (2008) 'Force and pressure measurement', in C. J. Payton and R. M. Bartlett (eds), *Biomechanical Evaluation of Movement in Sport and Exercise – The British Association of Sport and Exercise Sciences Guidelines*, pp. 53–76. Oxon: Routledge.

Laughton, C. A., McClay-Davies, I. and Hamill, J. (2003) 'Effect of strike pattern and orthotic intervention on tibial shock during running', *Journal of Applied Biomechanics*, 19: 153–168.

Linthorne, N. P. (2001) 'Analysis of standing vertical jumps using a force platform', *American Journal of Physics*, 69: 1198–1204.

McClay, I. S., Robinson, J. R., Andriacchi, T. P., Frederick, E. C., Gross, T., Martin, P. and Cavanagh, P. R. (1994) 'A profile of ground reaction forces in professional basketball', *Journal of Applied Biomechanics*, 10: 222–236.

McLean, B. (1994) 'The biomechanics of hurdling: force plate analysis to assess hurdling technique', *New Studies in Athletics*, 4: 55–58.

McNair, P. J., Prapavessis, H. and Callender, K. (2000) 'Decreasing landing forces: effect of instruction', *British Journal of Sports Medicine*, 34: 293–296.

Middleton, J., Sinclair, P. and Patton, R. (1999) 'Accuracy of centre of pressure measurement using a piezoelectric force platform', *Clinical Biomechanics*, 14: 357–360.

Mills, C., Yeadon, M. R. and Pain, M. T. (2010) 'Modifying landing mat material properties may decrease peak contact forces but increase forefoot forces in gymnastics landings', *Sports Biomechanics*, 9: 153–164.

Mornieux, G., Zameziati, K., Mutter, E., Bonnefoy, R. and Belli, A. (2006) 'A cycle ergometer mounted on a standard force platform for three-dimensional pedal forces measurement during cycling', *Journal of Biomechanics*, 39: 1296–1303.

Naemi, R., Chevalier, T. L., Healy, A. and Chockalingam, N. (2012) 'The effect of the use of a walkway and the choice of the foot on plantar pressure assessment when using pressure platforms', *The Foot*, 22: 100–104.

Nigg, B. M. (2006) 'Pressure distribution', in B. M. Nigg and W. Herzog (eds), *Biomechanics of the Musculo-skeletal System*, pp. 225–236. Chichester: Wiley Publishers.

Nigg, B. M. and Herzog, W. (2007) *Biomechanics of the Musculo-Skeletal System*, Chichester: Wiley Publishers.

Nilsson, J. and Thorstensson, A. (1989) 'Ground reaction forces at different speeds of human walking and running', *Acta Physiologica*, 136: 217–227.

Nunns, M. P. I., Dixon, S. J., Clarke, J. and Carre, M. (2016) 'Boot-insole effects on comfort and plantar loading at the heel and fifth metatarsal during running and turning in soccer', *Journal of Sports Sciences*, 34: 730–737.

Pataky, T. C. and LaFortune, M. (2015) 'Effects of footwear on driver clubhead speed on in amateur golfers: classical vs. Bayesian inference', *Footwear Science*, 7: S166–S167.

Rosenbaum, D. and Becker, H. P. (1997) 'Plantar pressure distribution and measurements: technical background and clinical applications', *Foot and Ankle Surgery*, 3: 1–14.

Rosenbaum, D., Westhues, M. and Bosch, K. (2013) 'Effect of gait speed changes on foot loading characteristics in children', *Gait and Posture*, 38: 1058–1060.

Schwartz, M. H., Rozumalski, A. and Trost, J. P. (2008) 'The effect of walking speed on the gait of typically developing children', *Journal of Biomechanics*, 41: 1639–1650.

Shorten, M. and Mientjes, M. (2011) 'The 'heel impact' force peak during running is neither 'heel' nor 'impact' and does not quantify shoe cushioning effects', *Footwear Science*, 3: 41–58.

Tessutti, V., Ribeiro, A. P., Trombini-Souza, F. and Sacco, I. C. N. (2012) 'Attenuation of foot pressure during running on four different surfaces: asphalt, concrete, rubber, and natural grass', *Journal of Sports Sciences*, 30: 1545–1550.

Woodward, C. M., James, M. K. and Messier, S. P. (1999) 'Computational methods used in determination of loading rate: experimental and clinical implications', *Journal of Applied Biomechanics*, 15: 404–417.

Yu, J., Sun, Y., Yang, C., Wang, D., Yin, K., Herzog, W. and Liu, Y. (2016) 'Biomechanical insights into differences between the mid-acceleration and maximum velocity phases of sprinting', *The Journal of Strength and Conditioning Research*, 30: 1906–1916.

CHAPTER 7

SURFACE ELECTROMYOGRAPHY

Adrian Burden

INTRODUCTION

Electromyography is a technique used for recording changes in the electrical potential of muscle fibres that are associated with their contraction. The fundamental unit of the neuromuscular system is the motor unit, which consists of the cell body and dendrites of a motor neuron, the multiple branches of its axon and the muscle fibres that it innervates. The number of muscle fibres belonging to a motor unit, known as the innervation number, can range from 5 to 1934 depending on the size of the muscle (Enoka, 2015). Detailed descriptions of the generation of an action potential by a motor neuron, its propagation along a muscle fibre and the processes that convert it into force within the fibre are beyond the scope of these guidelines and are more than adequately covered elsewhere (e.g. Enoka, 2015; Luttman, 1996). Knowledge of these processes is, however, paramount, as the biomechanist needs to understand the nature of the detected signal. In addition, and in agreement with Clarys and Cabri (1993), a detailed knowledge of musculoskeletal anatomy is also essential to ensure that electrodes are placed over the correct muscles.

The signal that is detected using surface electrodes is complex and is best understood by considering its underlying waveforms. The following terms originate from those introduced by Winter et al. (1980) and describe such waveforms as they develop from single action potentials into the detected signal. A muscle fibre action potential or a motor action potential (MAP) is the waveform detected from the depolarisation wave as it propagates in both directions along a muscle fibre. The waveform detected from the spatio-temporal summation of individual MAPs originating from muscle fibres belonging to a motor unit in the vicinity of a pair of recording electrodes is termed a motor unit action potential (MUAP). A motor unit action potential train (MUAPT) is the name given to the waveform detected from the repetitive sequence of MUAPs (i.e. repeated motor unit firing). Finally, the total signal seen at the electrodes is called the myoelectric signal (Winter et al., 1980), electromyographical signal (Basmajian and De Luca, 1985) or interference pattern (Enoka, 2015). This is the algebraic summation of all MUAPTs within the detection volume of the electrodes, and is represented schematically in Figure 7.1. The electromyographical signal when amplified and recorded, is termed the electromyogram (EMG).

Whilst these terms are useful in understanding the complex nature of the raw EMG, it is obvious that a single MAP could not be recorded using even the smallest surface electrodes. Individual MUAPs can, however, be identified from EMGs recorded using indwelling electrodes by a process known as decomposition (e.g. for a review, see Stashuk, 2001). More recently, individual MUAPs have been identified using a linear array or matrix of tiny (typically 1 mm diameter) surface electrodes (e.g. Merletti et al., 2001; Nawab et al., 2010; Kilby and Prasad, 2016). Unlike these electrodes, those that are used in the recording of EMGs from relatively large, superficial muscles during dynamic activities in sport and exercise are typically greater than 1 cm in diameter (circular electrodes) or length (rectangular electrodes). In addition, the muscle fibres of different motor units tend to overlap spatially (Cram and Kasman, 1998), which means that a given region of muscle could contain fibres from between 20 and 50 different motor units (Enoka, 2008). Thus, larger surface electrodes have the potential to detect MUAPTs from many different motor units, as depicted in Figure 7.1.

The preceding discussion presents the oversimplified view that only the number of active motor units and their firing rates can affect the EMG. In reality, a host of other anatomical, physiological and technical factors also have the

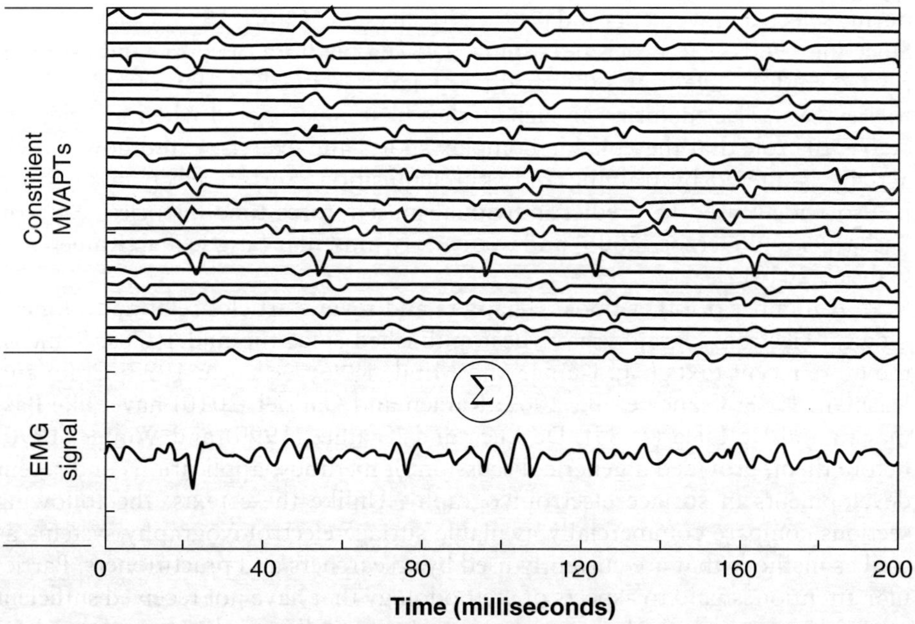

Figure 7.1 An EMG signal formed by adding (superimposing) 25 mathematically generated motor unit action potential trains (from Basmajian and De Luca, 1985).

potential to influence the electromyographical signal (e.g. Basmajian and De Luca, 1985; De Luca and Knaflitz, 1990; De Luca, 1997; Kamen and Gabriel, 2010). Such factors were grouped by De Luca (1997) into causative, intermediate and deterministic.

The causative factors have a fundamental influence on the EMG and were further divided into extrinsic and intrinsic groups. Extrinsic factors include, for example, the area and shape of the electrodes, the distance between them, and their location and orientation. Intrinsic factors include, for example, fibre type composition, muscle fibre diameter, their depth and location with regard to the electrodes, and the amount of tissue between the surface of the electrodes and the muscle; in addition to the number of active motor units. The intermediate factors, which are influenced by one or more of the causative factors, include the detection volume of the electrodes, their filtering effect, and cross-talk from neighbouring muscles. Finally, the deterministic factors, which are influenced by the intermediate factors and have a direct influence on the recorded EMG, include the amplitude, duration and shape of the MUAPs. The biomechanist has little influence over the intrinsic (causative) factors that can affect the EMG. However, the extrinsic (causative) factors are affected both by the techniques used by the biomechanist and their choice of equipment.

Basmajian and De Luca (1985), Cram and Kasman (1998) and Clarys (2000) have all previously documented the birth of electromyography in the eighteenth century and its development into the twentieth century. In general, electromyography has developed in two directions. Clinical electromyography is largely a diagnostic tool used to study the movement problems of patients with neuromuscular or orthopaedic impairment, whereas kinesiological electromyography is concerned with the study of muscular function and coordination during selected movements and postures (Clarys and Cabri, 1993; Clarys, 2000). Since the 1960s (e.g. Broer and Houtz, 1967) a plethora of sport- and exercise-related studies have been published (see Clarys and Cabri, 1993; Clarys, 2000 for reviews). The majority of such studies have investigated when muscles are active, the role that they play in complex sports and exercises, and how muscle activity is altered by training and skill acquisition. Surface electromyography is also widely used to predict individual muscle force-time histories (e.g. von Tscharner and Herzog, 2006) and to quantify muscular fatigue (e.g. Kamen and Gabriel, 2010).

A number of other books, chapters and review articles relating to kinesiological electromyography have been published since the mid-1970s. Some of the more recent texts (e.g. Kumar and Mital, 1996; De Luca, 1997; Cram and Kasman, 1998; Clancy et al., 2002; Kamen and Gabriel, 2010) have, like Basmajian and De Luca (1985), De Luca and Knaflitz (1990) and Winter (1990) before them, provided a generic discussion of methods, applications and recent developments in surface electromyography. Unlike these texts, the following sections compare commercially available surface electromyography systems as well as methods that are currently used by researchers and practitioners. Particular attention is paid to aspects of methodology that have not received sufficient attention previously, including reduction of cross-talk and the normalisation of processed EMGs. Recommendations are made based on information from the texts listed previously, other scientific literature, and both the International Society of Electrophysiology and Kinesiology (ISEK) (Winter et al., 1980; Merletti, 1999) and the SENIAM Project (Hermens et al., 1999); as well as the author's own experiences.

EQUIPMENT CONSIDERATIONS

It is generally accepted that the peak amplitude of the raw EMG recorded using surface electromyography does not exceed 5 mV and that its frequency spectrum is between 0–1000 Hz (e.g. Winter, 1990); with most of the usable energy limited to below 500 Hz and the dominant energy between 50–150 Hz (De Luca, 2002). When detecting and recording EMGs a major concern should be that the fidelity of the signal is maximised (De Luca, 1997). This is partly achieved by maximising the signal-to-noise ratio (i.e. the ratio of the energy in the electromyographical signal to that in the noise). Noise can be considered as any signals that are not part of the electromyographical signal and can include movement artefacts, detection of the electrocardiogram, ambient noise from other machinery, and inherent noise in the recording equipment. Details of these, and other, sources of noise are beyond the scope of these guidelines and are adequately covered elsewhere (e.g. Örtengren, 1996; Cram and Kasman, 1998; Kamen and Gabriel, 2010). In particular, Clancy et al. (2002) and Kamen and Gabriel (2010) provide very useful recommendations for reducing noise from a variety of sources. Maximising the fidelity of the EMG is also achieved by minimising the distortion (i.e. alteration of the frequency components of the signal) that it receives during detection, amplification and recording (De Luca, 2002). Both the equipment and procedures used to detect and record EMGs have a major influence on their fidelity, and should be given careful consideration.

Most commercially available electromyographical systems can be classified as either hard-wired, telemetry or data logger systems, with some companies offering more than one of these options (see Table 7.1).

A data logger or telemetry system is necessary if data are to be collected away from the main recording apparatus; however data loggers, typically do not allow on-line viewing of EMGs as they are being recorded and telemetry systems can be prone to ambient noise and cannot be used in areas with radiated electrical activity. Hard-wired systems do not suffer from these limitations, but obviously preclude data collection outside of the vicinity of the recording apparatus. The fidelity of the recorded electromyogram is dependent on these characteristics and, as such, should be a major consideration when purchasing a system. Most commonly, two electrodes are used to detect the electromyographical signal, i.e. a bipolar configuration. At the heart of the electromyographical system a differential amplifier subtracts the signal detected by one electrode from that detected by the other. Whilst a detailed explanation of amplifier characteristics is beyond the scope of these guidelines (for details see Basmajian and De Luca, 1985; Winter, 1990; Cram and Kasman, 1998; Clancy et al., 2002; Kamen and Gabriel, 2010) the important ones

Table 7.1 Summary of commercially available electromyography systems.

Company	*Type*	*Channels*	*Electrode type*
B&L Engineering www.bleng.com	Hard-wired	6	Circular fixed inter-electrode distances, and snap connection
Bioelettronica www.otbioelettronica.it	Hard-wired, telemetry and data logger	Up to 400	Snap connection, and 4–64 electrode arrays
Biometrics www.biometricsltd.com	Hard-wired, telemetry and data logger	8,16,24	Circular fixed inter-electrode distance, and snap connection
Biopac www.biopac.com	Hard-wired, telemetry and data logger	16	Circular fixed inter-electrode distance, and snap connection
Bortec www.bortec.ca	Hard-wired	8,16	Circular fixed inter-electrode distance, and snap connection
BTS Bioengineering www.btsbioengineering.com	Telemetry	16,20	Snap connection
Cadwell www.cadwell.com	Hard-wired	2,4,12	Snap connection
Cometa www.cometasystems.com	Telemetry and data logger	8,16,32	Snap connection, waterproof version
Delsys www.delsys.com	Hard-wired Telemetry and data logger	4,8,16 16	Bar fixed inter-electrode distance Bar fixed inter-electrode distance
Iworx www.iworx.com	Hard-wired	4,8	Snap connection
MIE www.mie-uk.com	Telemetry and data logger	16	
Mega www.megaemg.com	Telemetry and data logger	4,8,16	Snap connection
Motion Lab Systems www.motion-labs.com	Hard-wired and telemetry	6,8,10,16	Circular fixed inter-electrode distance, and snap connection
Noraxon www.noraxon.com	Telemetry and data logger	4,8,16	Snap connection, with option for fixed inter-electrode distance
zebris Medical GmbH www.zebris.de	Telemetry	8	Snap connection

are listed here, together with recommended requirements; which are generally agreed upon by De Luca (1997), the SENIAM Project (Merletti et al., 1999a) and Kamen and Gabriel (2010):

- Input Impedance (>100 MΩ)
- Common Mode Rejection Ratio (CMRR) (>80 dB [10 000])
- Input Referred Noise (<1–2 μV rms)
- Bandwidth (20–500 Hz)
- Gain (variable between 100 and 10,000; see Sampling)

Whilst the requirements of amplifiers are generally agreed on by biomechanists, the configuration of electrodes and the material from which they are made are not. The SENIAM Project (Freriks et al., 1999) prefers pre-gelled Ag/AgCl electrodes that are circular with a diameter of 10 mm and a centre-to-centre distance of 20 mm; such as that used by Bortec. In contrast, De Luca (1997) recommends silver bar electrodes that are 10 mm long, 1 mm wide, have a distance of 10 mm between them and are attached without the use of a gel; such as that used by Delsys. Considering a single MAP, the signal that passes beneath each electrode is biphasic, but their subtraction by a differential amplifier results in a triphasic waveform (e.g. Winter, 1990; Kamen and Gabriel, 2010). If the inter-electrode distance is increased the duration of the resulting triphasic waveform and its amplitude are increased, and vice versa (e.g. Kamen and Gabriel, 2010). This results in a lower bandwidth of the detected electromyographical signal. Assuming the conduction velocity of the depolarisation wave along a muscle fibre to be 4 m s^{-1}, an inter-electrode distance of at least 1 cm will result in a bandwidth that contains the full frequency spectrum of the raw EMG (Basmajian and De Luca, 1985).

DATA COLLECTION PROCEDURES

Electrode configuration, location and orientation

Assuming that a bipolar configuration is used, in order to maximise the amplitude of the detected signal the two electrodes should not be placed either side of and at the same distance from a motor point (i.e. a point on the skin where the lowest electrical stimulation will result in a minimal muscle twitch, Kamen and Gabriel, 2010). As differential amplifiers subtract the signals detected by the electrodes, locating electrodes either side of a motor point will lead to the cancellation of symmetrical action potentials that are travelling in opposite directions from the neuromuscular junction and that reach the electrodes at approximately the same time (e.g. Merletti et al., 2001). This is illustrated in Figure 7.2 by using the concept of single MAP detected by a pair of electrodes from a linear array (electrode pair and EMG number 2). However, if both electrodes are placed to one side of a motor point the signal is not cancelled to the same extent, as one electrode detects the MAP slightly earlier than the other (e.g. electrode pair and EMG number 3 in Figure 7.2).

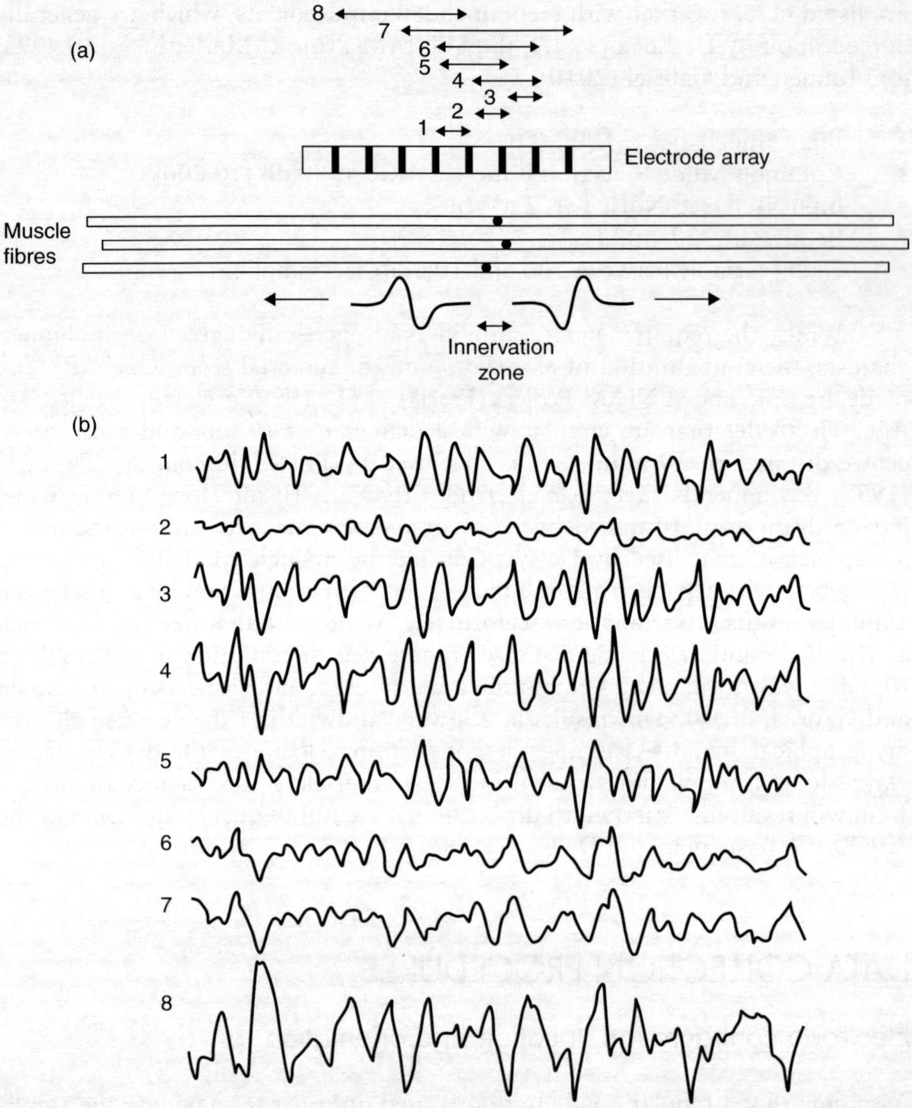

Figure 7.2 The influence of electrode location on EMG amplitude. (a) Eight electrodes arranged in an array, with a 10 mm spacing between each electrode. The lines (numbered 1 to 8) above the array indicate the different combinations of electrodes that were used to make bi-polar recordings. Inter-electrode distances are 10 mm for pairs 1, 2 and 3; 20 mm for pairs 4 and 5; 30 mm for pair 6; 40 mm for pair 8; and 50 mm for pair 7. (b) EMGs recorded using the array shown in (a) when placed on the skin overlying the biceps brachii at 70 per cent of MVC (adapted by Enoka, 2008 from Merletti et al., 2001).

Based on the preceding, it is generally agreed that electrodes should be located between a motor point and a tendon (e.g. De Luca, 1997; Freriks et al., 1999; Kamen and Gabriel, 2010). Figure 7.2 also illustrates the use of an array of electrodes, rather than a single pair, to locate motor points (Merletti et al., 2001; 2003). In the absence of a linear electrode array or a stimulator to detect the location of motor points, electrodes can be placed in the centre of the belly of the muscle whilst under contraction (Clarys and Cabri, 1993). However, it should be recognised that this location could coincide with that of a motor point.

Recommended guidelines for locating electrode placement sites over specific muscles also continue to be published (e.g. Zipp, 1982; Cram et al., 1998; Freriks et al., 1999). Virtually all major, superficial muscles are now covered by such guidelines, which typically locate electrodes about a point that is a specific distance along a line measured between two anatomical landmarks. Clarys and Cabri (1993) warned against using such recommendations for anything other than isometric contractions. If they are used, it should not be as a substitute for a good knowledge of musculoskeletal anatomy and the correct location should always be confirmed using voluntary activation of the muscle of interest and palpation. Following the location of an appropriate site, it is universally agreed that, if possible, the electrodes should be oriented along a line that is parallel to the direction of the underlying muscle fibres (e.g. De Luca, 1997; Freriks et al., 1999).

Skin preparation

The high input impedance that is offered by many of today's amplifiers has diminished the need to reduce the skin – electrode impedance to below levels around 10 kΩ. Some preparation of the skin (to below 55 kΩ) is, however, still necessary in order to obtain a better electrode – skin contact and to improve the fidelity of the recorded signal (e.g. Cram and Kasman, 1998; Freriks et al., 1999; Kamen and Gabriel, 2010). Typically, this involves cleaning the skin with soap and water and dry shaving it with a disposable razor. Additional rubbing with an alcohol-soaked pad and then allowing the alcohol to vaporise can be used to further reduce impedance, although this should be avoided when using participants with fair or sensitive skin.

In addition to the recording electrodes, differential amplifiers require the use of a reference, or ground, electrode that must be attached to electrically neutral tissue (e.g. a bony landmark). The degree of skin preparation given to the reference electrode site should be the same as that afforded to the muscle site. Some commercially available systems (e.g. Delsys) use a single, remote reference electrode for all muscle sites whilst others (e.g. MIE) use more local sites for each separate muscle that is being investigated. These latter systems usually have each pre-amplifier mounted on the press stud on the reference electrode and have relatively short leads between the pre-amplifier and the detecting electrodes. Most biomechanists also advise using an electrode gel or paste to facilitate detection of the underlying electromyographical signal (e.g. Cram and Kasman, 1998). This can be accomplished either through the use of pre-gelled electrodes (Freriks et al., 1999), which are often referred to as 'floating electrodes' (Kamen and Gabriel, 2010), or by applying a gel or paste to the skin or electrode prior to attachment (Clancy et al., 2002). Use of gel or paste is not always necessary when using so called active electrodes, i.e. those that are mounted onto the pre-amplifier (Kamen and Gabriel, 2010). Here, providing the skin is thoroughly cleaned the electrolytic medium is provided by the small amount of sweating that takes place when dry electrodes are applied to the skin (Cram and Kasman, 1998; Clancy et al., 2002; Kamen and Gabriel, 2010).

Cross-talk

Even if surface electrodes are placed close to the belly of the muscle it is possible that the detected signal may contain energy that emanates from other, more distant muscles. This form of noise, known as cross-talk, has been reported to be as high as 17 per cent of the signal from maximally activated nearby muscles, in both the lower (De Luca and Merletti, 1988) and upper leg (Koh and Grabiner, 1992). Despite research suggesting that cross-talk may not be as serious a problem as previously thought (e.g. Solomonow et al., 1994; Mogk and Keir, 2003), it is still an issue that biomechanists should address and attempt to reduce as much as possible. This is particularly pertinent when recording EMGs from muscles that are covered by thicker than normal amounts of subcutaneous fat, such as the gluteals and abdominals (Solomonow et al., 1994).

The presence of cross-talk can be estimated subjectively using functional tests (also known as manual muscle tests), which involve getting the participant to contract muscles that are adjacent to the one under investigation, without activating the one of interest. The detection of a signal from electrodes overlying the muscle of interest is, therefore, an indication of cross-talk. Whilst this technique has been found to provide an indication of the presence of cross-talk by the author, it is susceptible to two main limitations of which the biomechanist should be aware. The participant may not be able to activate nearby muscles, as pointed out by De Luca (1997); or the muscle of interest may be (moderately) activated in the process of trying to contract adjacent ones.

Cross-talk has traditionally been quantified either by comparing the amplitude of EMGs from a muscle with and without cross-talk (e.g. Koh and Grabiner, 1992), or by cross-correlating the signal from different muscles with cross-talk (e.g. Winter et al., 1994). Cross-correlation is a statistical technique that essentially quantifies the magnitude of any common component that is present in two separate signals i.e. from the muscle of interest and an adjacent one suspected of contributing to cross-talk; with a higher correlation indicating a stronger presence of cross-talk. Despite being criticised by De Luca (1997) it is still used by researchers to detect cross-talk (e.g. Mogk and Keir, 2003). Two further techniques for detecting the presence of cross-talk have been proposed by De Luca (1997). One involves comparing the frequency spectrum (see Processing EMGs in the Frequency Domain) of an EMG signal that is suspected of containing cross-talk with that of one believed to be primarily from the muscle of interest. The signal from a muscle further away from the electrodes would experience greater spatial (low pass) filtering than that from a muscle located directly below the electrodes. Thus, an EMG signal suspected of containing cross-talk would be expected to have a frequency spectrum that is more compressed towards the lower frequencies. The other technique involves using surface electrodes on muscles that are adjacent to the one of interest and wire electrodes in deep muscles, to monitor them for (lack of) activity. This approach is cumbersome and uses wire-electrodes, which may not be available.

Decreasing the size of the electrodes or the spacing between them reduces the chances of recording cross-talk (e.g. Koh and Grabiner, 1993; Winter et al., 1994; DeLuca et al., 2012). However, the use of excessively small inter-electrode distances will alter the bandwidth of the signal and may, therefore, compromise its fidelity. In addition, altering the size and spacing of the electrodes is often not possible due to the standard configuration used by some manufacturers (e.g. Bortec and Delsys). Research has shown that the most effective way of reducing cross-talk to almost negligible levels is to use the 'double differential technique' (De Luca and Merletti, 1988; Koh and Grabiner, 1993; DeLuca et al., 2012). Prior to this section, the discussion of equipment considerations and data collection procedures has referred to the use of a single pair of electrodes per muscle site (in addition to the reference electrode). In the vast majority of commercially available recording systems the electrodes are attached to a single differential amplifier (often referred to simply as a differential amplifier) that calculates the difference between the signals detected by each electrode overlying the muscle of interest. The amplifier used in the double differential technique has three, rather than two, detecting electrodes that are equally spaced apart, which calculates the difference between the signals detected by electrodes one and two, and electrodes two and three. These two new (single differentiated) signals are then further differentiated (double differentiation) by the amplifier (for further details refer to De Luca, 1997). This procedure works by significantly decreasing the detection volume of the three electrodes, and thereby filtering out signals from further away (De Luca, 1997). To the author's knowledge the double differential technique is only provided by Delsys. It is envisaged that other manufacturers will follow suit in the future and the double differential amplifier will be an option in all major recording systems. Recently, a blind signal separation (BSS) algorithm that aims to separate independent EMGs from measured signals has been successful in reducing cross-talk in forearm muscles (e.g. Kong et al., 2010).

Sampling

The Nyquist theorem dictates that electromyographical signals which are detected using surface electrodes should be sampled at a minimum of 1000 Hz (ideally 2000 Hz) to avoid aliasing (i.e. loss of information from the signal). Sampling of the signal into a PC should also use an analogue-to-digital converter (ADC) that has at least 12 bits (ideally 16 bits) to ensure that as small a change in muscle activity as possible is able to be detected by the system. Prior to recording EMGs caution should be exercised over selection of the gain of the pre-amplifier/amplifier. Of primary concern is that too high a gain is not chosen, which may result in the amplified signal exceeding the output voltage of the system (typically ±5 V). This is evidenced by 'clipping' (i.e. a raw signal that has had its negative and positive peaks chopped off at a specific amplitude), which often occurs when a very high gain (e.g. 10,000) is used for a muscle that typically emits a large signal (e.g. tibialis anterior). Care must also be taken not to use too low a gain (e.g. 100), as this may result in only a small part of the output voltage being utilised and hence small changes in signal amplitude not being quantified; particularly if a low resolution ADC is used (i.e. 12 bits or less). Ideally, a gain should be selected that utilises as much of the output voltage of the system as possible, whilst avoiding clipping. This typically ranges between 1000 and 5000 for superficial muscles in sport and exercise applications when using an output voltage of ±5 V. Most systems include pre-amplifiers with different gains or a choice of gain settings on the main amplifier to enable selection of an appropriate value. The reader is referred to Merletti et al. (1999a), Clancy et al. (2002) and De Luca (2003) for a more in depth discussion of the issues surrounding sampling of electromyographical signals.

PROCESSING, ANALYSING AND PRESENTING ELECTROMYOGRAMS

Processing EMGs in the time domain

Raw EMGs have been processed in numerous ways, particularly since the advent of computers (see Basmajian and De Luca, 1985; Winter, 1990). Today, raw EMGs are processed in the time domain almost exclusively using either the Average Rectified EMG, Root Mean Square or Linear Envelope. All of which provide an estimate of the amplitude of the raw EMG in μV or mV. The Average Rectified EMG (AREMG, e.g. Winter et al., 1994) has also been termed the Average Rectified Value (ARV, e.g. Merletti et al., 1999a), Mean Absolute Value (MAV, e.g. Clancy et al., 2002), and the Mean Amplitude Value (MAV, e.g. Merletti et al., 1999a); but will be referred to as the ARV for the remainder of this chapter. Calculation of the ARV involves first either removing all of the negative phases of the raw EMG (half-wave rectification) or reversing them (full wave rectification). The latter option is preferred, and almost exclusively performed today, because it retains all of the energy of the signal (Basmajian and De Luca, 1985). The integral of the rectified EMG is then calculated over a specific time period, or window (T), and the resulting integrated EMG is finally divided by T to form the ARV (see equation 7.1). The integrated EMG (reported in μV s or mV s) was a common processing method in its own right from the latter half of the twentieth century (see Winter, 1990). However, despite the ARV being similar to and providing no additional information to the integrated EMG the latter method is mainly used to provide an estimate of the total amount of activity.

$$\mathrm{ARV} = \frac{1}{T}\int_0^T |\mathrm{X}(t)|\,dt \qquad (7.1)$$

where:

X(t) is the EMG signal,

T is the time over which the ARV is calculated.

The Root Mean Square EMG (RMS) is the square root of the average power of the raw EMG calculated over a specific time period, or window (T) (see equation 7.2).

$$\text{RMS} = \sqrt{\frac{1}{T}\int_0^T X^2(t)dt} \tag{7.2}$$

Both the ARV and RMS are recognised as appropriate processing methods by De Luca's group (e.g. Basmajian and De Luca, 1985; De Luca and Knaflitz, 1990), the SENIAM Project (Merletti et al., 1999a), as well as other authors (e.g. Clancy et al., 2002). The same authors also prefer the RMS over the ARV, albeit for different reasons. De Luca (1997) states that the ARV is a measure of the area under the rectified EMG and thus has no specific physical meaning. In comparison, the RMS is a measure of the power of the signal and thereby has a clear physical meaning. Merletti et al. (1999a) claim that for a stationary signal, which may occur during a low level isometric contraction, the RMS shows less variability than the ARV when calculated over successive time windows. Thus, the RMS has the potential to detect signal changes that could be masked by the greater variability of the ARV.

Rather than using a single calculation of the RMS or ARV, the raw EMG is often processed by making successive calculations throughout its duration. The resulting series of values form a type of moving average (for further details see Winter et al., 1980 or Basmajian and De Luca, 1985). Regardless of which method is used, the biomechanist also needs to decide on the duration (or width) of the successive time windows (T). Basmajian and De Luca (1985) recommend choosing a width between 100 and 200 ms. Merletti et al. (1999a) further recommend that for a sustained isometric contraction, durations of 0.25–22 s are acceptable, with shorter widths (0.25–20.5 s) used for contraction levels above 50 per cent MVC and longer ones used (1–2 s) for lower level contractions. The smaller the width of the time window the less smooth the resulting curve will be (Basmajian and De Luca, 1985). Consequently, EMG variability over successive windows has been shown to decrease in an exponential fashion as the width of the processing window increases in submaximal (Norman et al., 1978) and maximal isometric contractions (e.g. Buckthorpe et al., 2012).

The moving average approach (using either RMS or ARV) described earlier is also used to analyse raw EMGs recorded from dynamic contractions, where the aim is often to detect (sometimes rapid) changes in muscle activity. Selection of short duration window widths (e.g. 10–50 ms) may allow the detection of rapid alterations in activity, but the resulting curve will still resemble the rectified EMG. Thus, peak amplitudes from repetitions of the same task will remain highly variable. Adoption of longer widths (e.g. 100–200 ms) will reduce the variability of peak amplitudes, but the resulting curve will lose the trend of the underlying EMG. As such, rapid changes in muscle activity may go undetected. Overlapping the windows by a progressively greater amount results in a curve that increasingly follows the trend of the underlying rectified EMG, but without the variable peaks that are evident in the rectified EMG. Similar to previous research using isometric contractions, Burden et al. (2014) discovered that reliability of the RMS EMG from the gluteus maximus during the clam exercise improved as the processing window was lengthened. This effect was greater in the peak RMS than the mean RMS values. As a consequence, statistical significance and, to a lesser extent, effect size between two variations of the exercise was more sensitive to changes in window width in the peak RMS value. Thus, we recommend use of the mean, rather than the peak, RMS when comparing muscle activity between such conditions.

The Linear Envelope is also still a popular processing method, and is recommended by Merletti et al. (1999a) for use on EMGs from dynamic contractions. Similar to the moving average, this involves smoothing the rectified EMG with a low pass filter (for further details see Winter et al., 1980, or Winter, 1990), and also results in a curve that follows the trend of the EMG.

When using the Linear Envelope, the type, order and cut-off frequency of the filter need to be decided on. Traditionally, a second order Butterworth filter has been applied (e.g. Winter et al., 1980) with a cut-off frequency that has ranged between 3 and 80 Hz (Gabel and Brand, 1994). Deciding on the cut-off frequency is similar to choosing the width and amount of overlap of the time window when using a moving average. Use of a low cut-off frequency will result in a very smooth curve, which will be unable to detect rapid changes in activation. Conversely, choice of a higher frequency will closely follow rapid changes in activity, but will still bear the peaks that characterise the rectified EMG. For further details of amplitude processing methods and their effect on the electromyographical signal the reader should refer to Clancy et al. (2002).

Threshold analysis

Following processing the EMG is often used to estimate when a muscle is active (i.e. on) or inactive (i.e. off). Typically, in order to determine the amplitude threshold at which the muscle is considered to be active, the baseline EMG (or noise) is treated as a stochastic variable. The mean of this baseline is, for example, calculated over 50 ms and the muscle is deemed to be active when the EMG amplitude exceeds three standard deviations above the mean baseline activity for 25 ms or more (Di Fabio, 1987). Numerous variations of these parameters have been proposed, and reviewed by Hodges and Bui (1996), Allison (2003) and Morey-Klapsing et al. (2004). In agreement with Di Fabio (1987), Hodges and Bui (1996) discovered that a sample width of 25 ms more accurately identified the onset of muscle activity, in relation to a visually determined value, than either 10 ms or 50 ms. They also reported that use of three standard deviations above mean baseline activity increased the risk of failing to identify EMG onset when it occurred (i.e. a Type II error), and delayed onset determination in relation to using two standard deviations. Following on from the discussion in the previous section, the degree of smoothing to which the EMG is subjected to during (time domain) processing will also affect the correct identification of EMG onset (Hodges and Bui, 1996; Allison, 2003). Use of a Linear Envelope with a cut-off frequency of 50 Hz was discovered by Hodges and Bui (1996) to result in a more accurate determination of muscle activity onset than either 10 Hz or 500 Hz.

A more conservative threshold of two standard deviations above the mean has also been included in previous recommendations (e.g. De Luca, 1997). Furthermore, for the muscle of interest to be considered to be either on or off the amplitude should exceed or fall below the threshold for a period that is greater than the physiological conditions that limit the resolution of this on/off time (i.e. the time taken for the signal to reach the electrodes from the innervation zone) (Allison, 2003). For most applications this should be at least twice this physiologically limited resolution, or 20 ms (De Luca, 1997). However, it must also be appreciated that the success of the chosen parameters in accurately and reliably detecting the onset of muscle activity will be affected by both the signal-to-noise ratio of the EMG (i.e. clean vs. noisy baseline) and the rate of increase in its amplitude during the specific task (i.e. slow vs. fast movements) (for further details see Allison, 2003). Thus, most authors (e.g. Di Fabio, 1987; Hodges and Bui, 1996; Allison, 2003) agree that whatever the chosen method it should be supplemented by repeated visual verification.

Ensemble averaging EMGs

It is often desirable to present and analyse the average (mean) EMG patterns from a number of trials, particularly for cyclic events such as walking, running and cycling. Such a pattern of processed EMGs is termed an ensemble average (e.g. Winter et al., 1980). The first stage in producing an ensemble average is to decide on a start and an end point for each trial (e.g. heel contact to heel contact of the same foot for a walking stride). The duration of each trial is then converted to a new standardised time-base (e.g. the percentage of the stride time). The process is often referred to as 'temporal normalisation' and should not be confused with amplitude normalisation, which is described in the next section. Next, data from each trial are interpolated at the same points on the new time-base, for example, 2 or 5 per cent intervals. This can be done using linear interpolation or by fitting a cubic spline to the data points. Finally, data from all trials are averaged at each of the points on the new time-base. The resulting ensemble average is typically displayed along with an appropriate measure of the variability between trials (e.g. ±1 standard deviation).

Ensemble averages can be produced for the same individual over a number of strides (i.e. intra-individual) as described earlier, and also for a number of individuals. An inter-individual ensemble average is created similarly by calculating the mean of the intra-individual ensemble averages at each of the points on the time-base.

Assessment of intra-individual and inter-individual variability of EMG patterns has traditionally been accomplished using either the coefficient of variation or the variance ratio (VR in equation 7.3). The coefficient of variation was promoted by Winter and colleagues throughout the 1980s and 1990s (e.g. Yang and Winter, 1984), whilst the variance ratio has sporadically been used to measure the variability of EMG patterns (e.g. Burden et al., 2003) after initially being devised by Hershler and Milner (1978). Unlike the variance ratio, the coefficient of variation is essentially the ratio of the standard deviation of the EMG patterns to the mean pattern. Therefore, the magnitude of the mean has the potential to either elevate or reduce the value of the coefficient. As a consequence, the coefficient of variation is sensitive to both the amount of smoothing applied to the EMG during processing and the number of cycles included in the ensemble average (Gabel and Brand, 1994). For these reasons the coefficient variation should not be used to compare the variability of EMG patterns from different muscles; and certainly not to compare the variability of EMG patterns with those of other biomechanical parameters. It is recommended that the variance ratio be used to assess the variability of EMG patterns, particularly when comparisons of variability are to be made across muscles, individuals, etc.

$$\mathrm{VR} = \frac{\sum_{i-1}^{k}\sum_{j=1}^{n}(X_{ij} - \overline{X}_i)^2 \,/\, k(n-1)}{\sum_{i-1}^{k}\sum_{j=1}^{n}(X_{ij} - \overline{X})^2 \,/\, (kn-1)} \tag{7.3}$$

where:

k is the number of time intervals within the cycle,
n is the number of cycles (for intra-individual variability) or participants (for inter-individual variability),
X_{ij} is the EMG value at the ith time for the jth cycle or participant,
$\overline{X}_i$ is the mean of the EMG values at the ith time over the jth cycle or participant.

$\overline{X}$ is the mean of the EMG values i.e. $\overline{X} = \frac{1}{k}\sum_{i=1}^{k} X_i$

NORMALISING EMGS

Relocation of electrodes over the same muscle on subsequent occasions will invariably result in the detection of different motor units. The skin-electrode impedance will also differ between sessions, regardless of how well skin preparation techniques are adhered to, which will affect the shape of the underlying MUAPs. These and other factors that were previously outlined in the Introduction Section will, therefore, affect the amplitude of the processed EMG. As such, the amplitude of EMGs recorded from the same muscle on different occasions, as well as from different muscles and different individuals should not be compared directly, even if they have been processed using the same method. This problem may be solved by (amplitude) normalisation of EMGs after they have been processed. Traditionally, this has involved expressing each data point of the processed EMG from the specific task as a proportion or a percentage of the peak EMG from an isometric maximal voluntary contraction (MVC) of the same muscle that has been processed in the same way (see Burden, 2010, for a review of this and other normalisation methods). In addition to allowing the comparison of EMGs from different muscles, this normalisation method has the potential to reveal how active a muscle is during a specific task; by expressing it as a proportion or percentage of its maximal activation capacity. However, despite being recommended by ISEK (Merletti, 1999) the SENIAM project (Merletti et al., 1999b) and Burden (2010) this method has received several criticisms that biomechanists should be conscious of before using it.

Among others, Clarys (2000) recognised that this method can be limited by the poor reliability of EMGs recorded from MVCs. Because of this, the reportedly improved reliability of EMGs from sub-maximal contractions has resulted in these EMGs being used as the denominator in the normalisation equation (e.g. Yang and Winter, 1984). In particular, De Luca (1997) advocates the use of EMGs from contractions that are less than 80 per cent MVC in order to provide a more stable reference value. Sub-maximal contractions can also be used when normalising EMGs from individuals who are unable or unwilling to perform MVCs due to musculoskeletal pain or injury. For example, Healey et al. (2005) obtained a reference contraction of the lumbar erector spinae in individuals with chronic low back pain by holding a 4 kg bar at arm's length for 15 s. A further criticism of using EMGs from isometric MVCs centres on whether the maximal activation capacity of the muscle is actually elicited during the voluntary effort. Outputs from this method of normalisation which are in excess of unity (i.e. greater than 100 per cent) indicate that the activity of the muscle during a specific task can be greater than that recorded during the isometric MVC. For example, Jobe et al. (1984) reported that EMGs from the serratus anterior were 226 per cent of the EMG from a maximal muscle test during the acceleration phase of the overarm throw. Whilst research using the twitch interpolation technique has revealed that (most) individuals are able to maximally activate their muscles during isometric MVCs (e.g. Allen et al., 1995), the same authors do warn that the ability to consistently achieve maximal activation varied between individuals. A recent review by Ball and Scurr (2013) further questioned the use of EMGs from an isometric MVC to normalise EMGs from, particularly high-velocity, dynamic contractions (tasks). From a range of normalisation methods they (Ball and Scurr, 2010) reported that EMGs from maximal dynamic tasks such as sprinting and squat jumping generally exhibited the best reliability and greatest amplitude, and advised using these to normalise EMGs from dynamic activities. They (Ball and Scurr, 2013) further recommended that the normalisation method should use an, ideally familiar, action similar to the task under investigation.

Regardless of whether electromyographers use EMGs from isometric or dynamic MVCs it is strongly recommended that individuals receive extensive practice in performing them before EMGs from them are used for normalisation. Previously unrehearsed MVCs will result in torque or force, and hence muscle activity, that is far from maximal.

Amplitude normalisation can also be used to reduce inter-individual variability of EMGs recorded from the same task. It is now well established that dividing each data point within the task EMG by either the mean or the peak EMG from the same task EMG is the most effective way of improving group homogeneity (e.g. Yang and Winter, 1984; Burden et al., 2003). However, due to the nature of the denominator used in their normalisation equation, these methods should not be used to compare the amplitude of EMGs from different muscles and individuals (see Burden, 2010, for a review).

Processing EMGs in the frequency domain

Raw EMGs are processed in the frequency domain primarily to investigate changes in the signal that accompany muscular fatigue. It is now well established that fatigue is associated with a compression of the frequency spectrum towards the lower frequencies (see Figure 7.3), that occurs largely due to a decrease in the conduction velocity of action potentials and a consequent increase in the duration of MUAPs. The reader is referred to Basmajian and De Luca (1985) or De Luca (1997) for a more detailed explanation of these processes and the physiological mechanisms behind them.

Transformation of a raw EMG from the time domain to the frequency domain is typically achieved using a Fast Fourier Transform (FFT). Details of how the FFT decomposes the signal into its various frequency contributions are beyond the scope of these guidelines, and the reader is referred to Enoka (2015) or Giakas (2004) for further details. The FFT requires specification of the time window (or epoch) over which the spectrum is calculated and the shape of the window (e.g. rectangular or more complex such as Hanning). Merletti et al. (1999a) recommend a time window of between 0.5 and 1 s, but also reveal that the window shape has little effect on parameters used to represent the frequency spectrum (see below).

The output of the FFT is typically represented as the power spectrum density (PSD), which shows the relative magnitudes of the range of frequencies present in the raw signal (see bottom of Figure 7.3). One of two parameters is commonly obtained from the PSD in order to quantify it. The median frequency (MDF) is defined as the frequency that divides the PSD into equal halves, and the mean frequency (MNF) is calculated as the sum of the product of the individual frequencies and their own power divided by the total power (e.g. Finsterer, 2001). Whilst the MNF is less variable than the MDF, the latter is less sensitive to noise and more sensitive to spectral compression (e.g. De Luca and Knaflitz, 1990; Merletti et al., 1999a). As such, both SENIAM (Merletti et al., 1999a) and De Luca (e.g. De Luca, 1997) recommend use of the MDF as an indicator of muscular fatigue.

Regardless of the chosen parameter, it is typically obtained from consecutive time windows, to enable changes in the signal that occur as a consequence of fatigue to be monitored. Successive values from the contraction period are then analysed using (linear) regression; with the intercept of the regression line being the initial frequency and the gradient representing the fatigue rate (e.g. Ng et al., 1996).

In addition to fatigue, the frequency spectrum of the raw EMG is affected by a host of other factors (see Introduction section). Similar to analysis in the

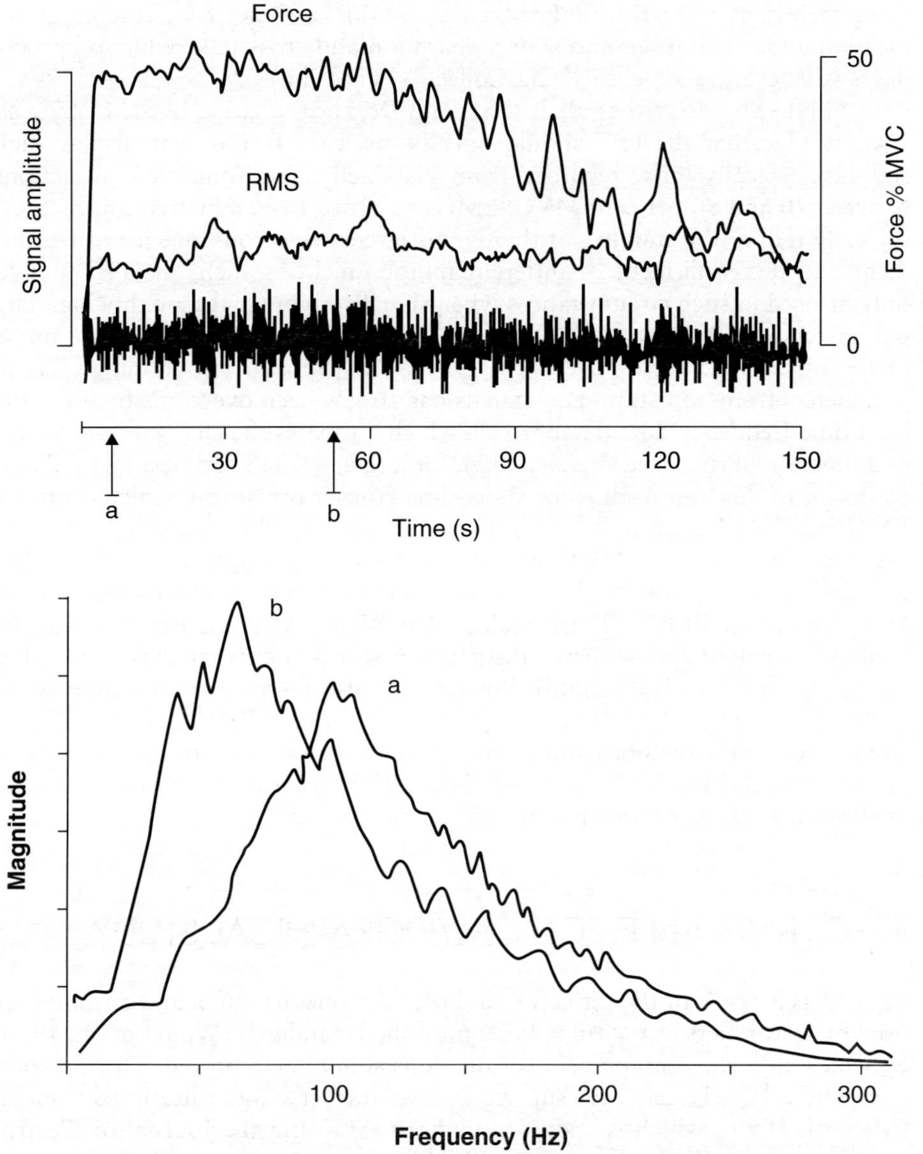

Figure 7.3 (Top) EMG signal amplitude and force during an attempted constant-force contraction of the first dorsal interosseus muscle. (Bottom) Power spectrum density of the EMG signal at the beginning (a) and at the end (b) of the constant force segment of the contraction (from Basmajian and De Luca, 1985).

time domain (see Normalising EMGs section) specific frequencies (e.g. MDF) cannot, therefore, be compared directly when they are calculated from EMGs recorded from different muscles or individuals, or from the same muscle when the electrodes have been re-applied. However, comparisons can be made between the gradient of the regression line in order to investigate differences in fatigue rates between different muscles, occasions or individuals (e.g. Ng et al., 1996).

Nevertheless, care should still be taken to use the same electrode configuration, location and orientation, and skin preparation and cross-talk reduction procedures whenever possible (see Data Collection Procedures).

SENIAM (Merletti et al., 1999a) and De Luca (e.g. De Luca, 1997) also acknowledge that the FFT should only be used on EMGs that display high stability; typically those recorded from sustained force isometric contractions between 20 and 80 per cent MVC. EMGs recorded from dynamic contractions typically reduce the stability of the signals largely as a consequence of recruitment and de-recruitment of different motor units. As such, the FFT should only be used in such circumstances when signal stability is reasonably high and parameters (i.e. MDF or MNF) should only be calculated at the same phase of repetitive cyclic events (De Luca, 1997). The problem of obtaining spectral parameters from non-stationary signals has largely been overcome by using the joint time-frequency domain approach which estimates the change in frequency as a function of time (see Giakas, 2004, for a review). The simplest method that conforms to this approach is the short-time Fourier transform, which splits the EMG into small continuous or overlapping time windows, applies a FFT to each and calculates the MDF or MNF as above (e.g. Englehart et al., 1999). Recently, more sophisticated methods of time-frequency domain analysis have been applied to EMGs. These include the Wigner-Ville distribution wavelet analysis and the Choi-Williams distribution (for a review of these and other methods, see Reaz et al., 2006). Readers are also referred to Englehart et al. (1999), Karlsson et al. (2000), Hostens et al. (2004) and Beck et al. (2005) for a comparison of the methods and performance of various time-frequency domain approaches, and von Tscharner and Herzog (2006) for applications of wavelet analysis in addition to the investigation of muscle fatigue.

REPORTING AN ELECTROMYOGRAPHICAL STUDY

Largely as a result of the different, and often erroneous, units and terminology used by researchers at the time, ISEK published standards (Winter et al., 1980) by which electromyographical investigations should be reported. These recommendations have largely been superseded by a further standardisation document (Merletti, 1999), which is endorsed by both ISEK and the Journal of Electromyography and Kinesiology. The SENIAM project also published guidelines for reporting EMGs recorded using surface electromyography (Merletti et al., 1999b) which are almost identical to those produced by ISEK. The following information is based largely on that provided by ISEK and the SENIAM project, and should be considered as the minimum detail required:

Equipment

- Type of system (e.g. hard-wired, telemetry)
- Manufacturer(s) and model of the system, or different components (e.g. electrodes and amplifier) of the system

- Material, size and shape of electrodes
- Type of amplifier (i.e. differential or double differential) and the following characteristics:
  - Common Mode Rejection Ratio (CMRR) (dB)
  - Input impedance (MΩ)
  - Gain
  - Input referred noise (either peak-to-peak or rms) (μV) or signal-to-noise ratio (dB)
  - Bandwidth (Hz), including details of the filter types (e.g. Chebyshev, Butterworth) and slopes of the cut-offs at each end of the bandwidth (dB/octave or dB/decade)
- Manufacturer, model and number of bits of the ADC used to sample data into the computer.

Data collection procedures

- Location of the electrodes (ideally) with respect to the motor point
- Orientation of the electrodes with respect to the underlying muscle fibres
- Inter-electrode distance
- Skin-preparation techniques, including details of any gel or paste used
- Procedures used to check for the presence of cross-talk and/or reduce the amount of cross-talk
- Sampling frequency (Hz).

Processing and analysing EMGs

- Amplitude processing
  - Method used to process the raw EMG (e.g. RMS, ARV, Linear Envelope)
  - For ARV and RMS: width of time window and (if applicable) overlap of time window
  - For Linear Envelope: type and order of filter, and cut-off frequency
- Method used to normalise processed EMGs (e.g. Mean Dynamic, Isometric MVC, Reference Contraction), if applicable. If the Isometric MVC or Reference Contraction method was used then details of how the participants were trained to perform these contractions, their position, stabilisation methods and joint angles of involved limbs should be included
- Threshold detection method (e.g. level and duration), if applicable
- Details of ensemble averaging, if applicable
- Frequency processing
  - Algorithm used (e.g. FFT, wavelet analysis)
  - Length and type of window used (e.g. rectangular, Hanning for FFT)
  - Choice of parameter (e.g. MDF, MNF).

REFERENCES

Allen, G. M., Gandevia, S. C. and McKenzie, D. K. (1995) 'Reliability of measurements of muscle strength and voluntary activation using twitch interpolation', *Muscle and Nerve*, 18: 593–600.

Allison, G. T. (2003) 'Trunk muscle onset detection technique for EMG signals with ECG artefact', *Journal of Electromyography and Kinesiology*, 13: 209–216.

Ball, N. and Scurr, J. (2010) 'An assessment of the reliability and standardisation of tests used to elicit reference muscle actions for electromyographical normalisation', *Journal of Electromyography and Kinesiology*, 20: 81–88.

Ball, N. and Scurr, J. (2013) 'Electromyography normalization methods for high-velocity muscle actions: review and recommendations', *Journal of Applied Biomechanics*, 29: 600–6008.

Basmajian, J. V. and De Luca, C. J. (1985) *Muscles Alive: Their Functions Revealed by Electromyography*, 5th edn. Baltimore: Williams and Wilkins.

Beck, T. W., Housh, T. J., Johnson, G. O., Weir, J. P., Cramer, J. T., Coburn, J. W. and Malek, M. H. (2005) 'Comparison of Fourier and wavelet transform procedures for examining the mechanomyographic and electromyographic frequency domain responses during fatiguing isokinetic muscle actions of the biceps brachii', *Journal of Electromyography and Kinesiology*, 15: 190–199.

Broer, R. and Houtz, J. (1967) *Patterns of Muscle Activity in Selected Sports Skills: An Electromyographic Study*, Springfield, Il: Charles, C. Thomas.

Buckthorpe, M. W., Hannah, R., Pain, M. T. G. and Folland, J. P. (2012) 'Reliability of neuromuscular measurements during explosive isometric contractions, with special reference to electromyography normalization techniques', *Muscle Nerve*, 46: 566–576.

Burden, A. M. (2010) 'How should we normalize electromyograms obtained from healthy participants? What we have learned from over 25 years of research', *Journal of Electromyography and Kinesiology*, 20: 1023–1035.

Burden, A. M., Lewis, S. E. and Willcox, E. (2014) 'The effect of manipulating root mean square window length and overlap on reliability, inter-individual variability, statistical significance and clinical relevance of electromyograms', *Manual Therapy*, 19: 595–601.

Burden, A. M., Trew, M. and Baltzopoulos, V. (2003) 'Normalisation of gait EMGs: a re-examination', *Journal of Electromyography and Kinesiology*, 13: 519–532.

Clancy, E. A., Morin, E. L. and Merletti, R. (2002) 'Sampling, noise-reduction and amplitude estimation issues in surface electromyography', *Journal of Electromyography and Kinesiology*, 12: 1–16.

Clarys, J. P. (2000) 'Electromyography in sports and occupational settings: an update of its limits and possibilities', *Ergonomics*, 43: 1750–1762.

Clarys, J. P. and Cabri, J. (1993) 'Electromyography and the study of sports movements: a review', *Journal of Sports Sciences*, 11: 379–448.

Cram, J. R. and Kasman, G. S. (1998) *Introduction to Surface Electromyography*, Gaithersberg, Maryland, MD: ASPEN.

Cram, J. R., Kasman, G. S. and Holtz, J. (1998) 'Atlas for electrode placement', in J. R. Cram and G. S. Kasman (eds), *Introduction to Surface Electromyography*, pp. 223–388. Gaithersberg, Maryland, MD: ASPEN.

De Luca, C. (1997) 'The use of surface electromyography in biomechanics', *Journal of Applied Biomechanics*, 13: 135–163.

De Luca, C. (2002) Surface electromyography: detection and recording. [WWW]. www.delsys.com/Attachments_pdf/WP_SEMGintro.pdf (accessed 22 June 2016).

De Luca, C. J. and Knaflitz, M. (1990) *Surface Electromyography: What's New?* Boston, MA: Neuromuscular Research Centre.

De Luca, C. J. and Merletti, R. (1988) 'Surface myoelectric signal cross-talk among muscles of the leg', *Electroencephalography and Clinical Neurophysiology*, 69: 568–575.

De Luca, G. (2003) Fundamental concepts in EMG signal acquisition (Revision 2.1). [WWW]. www.delsys.com/Attachments_pdf/WP_Sampling1-4.pdf (accessed 22 June 2016).

DeLuca, C. J., Kuznetsov, M, Gilmore, L. D. and Roy, S. H. (2012) 'Inter-electrode spacing of surface EMG sensors: reduction of crosstalk contamination during voluntary contractions', *Journal of Biomechanics*, 45: 555–561.

Di Fabio, R. P. (1987) 'Reliability of computerized surface electromyography for determining the onset of muscle activity', *Physical Therapy*, 67: 43–48.

Englehart, K., Hudgins, B., Parker, P. A. and Stevenson, M. (1999) 'Classification of the myoelectric signal using time-frequency based representations', *Medical Engineering and Physics, Special Issue: Intelligent Data Analysis in Electromyography and Electroneurography*, 21: 431–438.

Enoka, R. M. (2008) *Neuromechanics of Human Movement*, 4th edn. Champaign, IL: Human Kinetics.

Enoka, R. M. (2015) *Neuromechanics of Human Movement*, 5th edn. Champaign, IL: Human Kinetics.

Finsterer, J. (2001) 'EMG-interference pattern analysis', *Journal of Electromyography and Kinesiology*, 11: 231–246.

Freriks, B., Hermans, H. J., Disselhorst-King, C. and Rau, G. (1999) 'The recommendations for sensors and sensor placement procedures for surface elctromyography', in H. J. Hermens, B. Freriks, R. Merletti, D. Stegeman, J. Blok, G. Rau, C. Disselhorst- Klug and G. Hägg (eds), *European Recommendations for Surface Electromyography: Results of the SENIAM Project*, pp. 15–54. Enschede: Roessingh Research and Development.

Gabel, R. H. and Brand, R. A. (1994) 'The effects of signal conditioning on the statistical analyses of gait EMG', *Electroencephalography and Clinical Neurophysiology*, 93: 188–201.

Giakas, G. (2004) 'Power spectrum analysis and filtering', in N. Stergiou (ed.), *Innovative Analyses of Human Movement*, pp. 223–258. Champaign, IL: Human Kinetics.

Healey, E. L., Fowler, N. E., Burden, A. M. and McEwan, I. M. (2005) 'The influence of different unloading positions upon stature recovery and paraspinal muscle activity', *Clinical Biomechanics*, 20: 365–371.

Hermens, H. J., Freriks, B., Merletti, R., Stegeman, D., Blok, J., Rau, G., Disselhorst-Klug, C. and Hägg, G. (eds) (1999) *European Recommendations for Surface Electromyography: Results of the SENIAM Project*, Enschede: Roessingh Research and Development.

Hershler, C. and Milner, M. (1978) 'An optimality criterion for processing electromyographic (EMG) signals relating to human locomotion', *IEEE Transactions on Biomedical Engineering*, 25: 413–420.

Hodges, P. W. and Bui, B. H. (1996) 'A comparison of computer-based methods for the determination of onset of muscle contraction using electromyography', *Electroencephalography and Clinical Neurophysiology*, 101: 511–519.

Hostens, I., Seghers, J., Spaepen, A. and Ramon, H. (2004) 'Validation of the wavelet spectral estimation technique in biceps brachii and brachioradialis fatigue assessment during prolonged low-level static and dynamic contractions', *Journal of Electromyography and Kinesiology*, 14: 205–215.

Jobe, F. W., Radovich, D., Tibone, J. E. and Perry, J. (1984) 'An EMG analysis of the shoulder in pitching: a second report', *The American Journal of Sports Medicine*, 12: 218–220.

Kamen, G. And Gabriel, D. A. (2010) *Essentials of Electromyography*, Champaign, IL: Human Kinetics.

Karlsson, S, Yu, J. and Akay, M. (2000) 'Time-frequency analysis of myoelectric signals during dynamic contractions: a comparative study', *IEEE Transactions on Biomedical Engineering*, 47: 228–238.

Kilby, J. and Prasad, K. (2016) 'Multi-channel surface electromyography electrodes: a review', *IEEE Sensors Journal*, 14: 5510–5519.

Koh, T. J. and Grabiner, M. D. (1992) 'Cross-talk in surface electromyograms of human hamstring muscles', *Journal of Orthopaedic Research*, 10: 701–709.

Koh, T. J. and Grabiner, M. D. (1993) 'Evaluation of methods to minimize cross-talk in surface electromyography', *Journal of Biomechanics*, 26: 151–157.

Kong, Y. K, Hallbeck, M. S and Jung, M. C. (2010) 'Crosstalk effect on surface electromyogram of the forearm flexors during a static grip task', *Journal of Electromyography and Kinesiology*, 20: 1223–1229.

Kumar, S. and Mital, A. (eds) (1996) *Electromyography in Ergonomics*, London: Taylor and Francis.

Luttman, A. (1996) 'Physiological basis and concepts of electromyography', in S. Kumar and A. Mital (eds), *Electromyography in Ergonomics*, pp. 51–96. London: Taylor and Francis.

Merletti, R. (1999) 'Standards for reporting EMG data', *Journal of Electromyography and Kinesiology (all volumes)*.

Merletti, R., Farina, D., Hermens, H., Frericks, B. and Harlaar, J. (1999a) 'European recommendations for signal processing methods for surface electromyography', in H. J. Hermens, B. Freriks, R. Merletti, D. Stegeman, J. Blok, G. Rau, C. Disselhorst-Klug and G. Hägg (eds), *European Recommendations for Surface Electromyography: Results of the SENIAM Project*, pp. 57–68. Enschede: Roessingh Research and Development.

Merletti, R., Wallinga, W., Hermens, H. J. and Freriks, B. (1999b) 'Guidelines for reporting SEMG data', in H. J. Hermens, B. Freriks, R. Merletti, D. Stegeman, J. Blok, G. Rau, C. Disselhorst-Klug and G. Hägg (eds), *European Recommendations for Surface Electromyography: Results of the SENIAM Project*, pp. 103–105. Enschede: Roessingh Research and Development.

Merletti, R., Rainoldi, A. and Farina, D. (2001) 'Surface electromyography for noninvasive characterization of muscle', *Exercise and Sport Sciences Reviews*, 29: 20–25.

Merletti, R., Farina, D. and Gazzoni, M. (2003) 'The linear electrode array: a useful tool with many applications', *Journal of Electromyography and Kinesiology*, 13: 37–47.

Mogk, J. P. M. and Keir, P. J. (2003) 'Crosstalk in surface electromyography of the proximal forearm muscles during gripping tasks', *Journal of Electromyography and Kinesiology*, 13: 63–71.

Morey-Klapsing, G., Arampatzis, A. and Brüggemann, G. P. (2004) 'Choosing EMG parameters: comparison of different onset determination algorithms and EMG integrals in a joint stability study', *Clinical Biomechanics*, 19: 196–201.

Nawab, S. H., Chang, S. and De Luca, C. J. (2010) 'High yield decomposition of surface EMG signals', *Clinical Neurophysiology*, 121: 1602–1615.

Ng, J. K. F., Richardson, C. A., Kippers, V., Parnianpour, M. and Bui, B. H. (1996) 'Clinical applications of power spectral analysis of electromyographic investigations in muscle tension', *Manual Therapy*, 2: 99–103.

Norman, R. W., Nelson, R. C. and Cavanagh, P. R. (1978) 'Minimum sampling time required to extract stable information from digitized EMGs', in K. Asmussen and E. Jørgensen (eds.), *Biomechanics VI-A*, pp. 237–243. Baltimore, MD: University Park Press.

Örtengren, R. (1996) 'Noise and artefacts' in S. Kumar and A. Mital (eds), *Electromyography in Ergonomics*, pp. 97–108. London: Taylor and Francis.

Reaz, M. B. I., Hussain, M. S. and Mohd-Yasin, F. (2006) 'Techniques of EMG signal analysis: detection, processing, classification and applications', *Biological Procedures Online*, 8; 11–35.

Solomonow, M., Baratta, R., Bernardi, M., Zhou, B., Lu, Y., Zhu, M. and Acierno, S. (1994) 'Surface and wire EMG crosstalk in neighbouring muscles', *Journal of Electromyography and Kinesiology*, 4: 131–142.

Stashuk, D. (2001) 'EMG signal decomposition: how can it be accomplished and used?' *Journal of Electromyography and Kinesiology*, 11: 151–173.

von Tscharner, V and Herzog, W. (2006) 'EMG', in B. M. Nigg and W. Herzog (eds), *Biomechanics of the Musculo-Skeletal System*, 3rd edn., pp. 409–452. Chichester: Wiley.

Wakeling, J.M., Pascual, S.A., Nigg, B.M. and von Tscharner, V. (2001) 'Surface EMG shows distinct populations of muscle activity when measured during sustained submaximal exercise', European Journal of Applied Physiology, 86: 40–47.

Winter, D.A. (1990) Biomechanics and Motor Control of Human Movement, 2nd edn, New York: Wiley.

Winter, D.A., Rau, G., Kadefors, R., Broman, H. and De Luca, C.J. (1980) Units, Terms and Standards in the Reporting of EMG Research, USA: International Society of Electrophysiological Kinesiology.

Winter, D.A., Fuglevand, A.J. and Archer, S.E. (1994) 'Crosstalk in surface electromyography: theoretical and practical estimates', Journal of Electromyography and Kinesiology, 4: 5–26.

Yang, J.F. and Winter, D.A. (1984) 'Electromyographic amplitude normalization methods: improving their sensitivity as diagnostic tools in gait analysis', Archives of Physical Medicine and Rehabilitation, 65: 517–521.

Zipp, P. (1982) 'Recommendations for the standardization of lead positions in surface electromyography', European Journal of Applied Physiology and Occupational Physiology, 50: 41–54.

CHAPTER 8

ISOKINETIC DYNAMOMETRY

Vasilios Baltzopoulos

INTRODUCTION

Isokinetic dynamometry is the assessment of dynamic muscle strength, and function in general, by measuring the joint moment exerted during constant joint angular velocity movements. It has widespread applications in sport, exercise and pathological conditions for the measurement of muscle strength, delivery and assessment of training or rehabilitation programmes, prediction of performance and prevention of injuries. Research applications include examinations of the mechanics of muscles, tendons and joints, modelling and simulation, and many other related areas (Baltzopoulos and Brodie, 1989; Kellis and Baltzopoulos, 1995; Yeadon et al., 2006; Conceição et al., 2012). Isokinetic dynamometers are extremely useful and unique devices that allow the assessment of dynamic muscle and joint function under specific joint angular conditions. However, there are many factors, related mainly to the biomechanical and physiological complexities of the musculoskeletal system, that could affect the measurements. It is, therefore, important to understand the mechanical basis of the measurements and the basic biomechanics of joint motion to be able to use an isokinetic dynamometer properly.

Human movement is produced by the rotation of segments around the instantaneous axis of rotation of the joints. Mechanically, rotation is only possible with the application of a moment. Muscles generate muscle forces that are transmitted to the bones via tendons. As an example, we will consider a knee extension motion in the sagittal plane only, around an instantaneous axis of rotation that is perpendicular to the plane of the segment motion. Assuming that there is only one active muscle-tendon unit, the moment producing the rotation can be determined simply as the product of the muscle force and its moment arm, that is defined as the shortest distance between the muscle line of action and the joint axis of rotation (Figure 8.1).

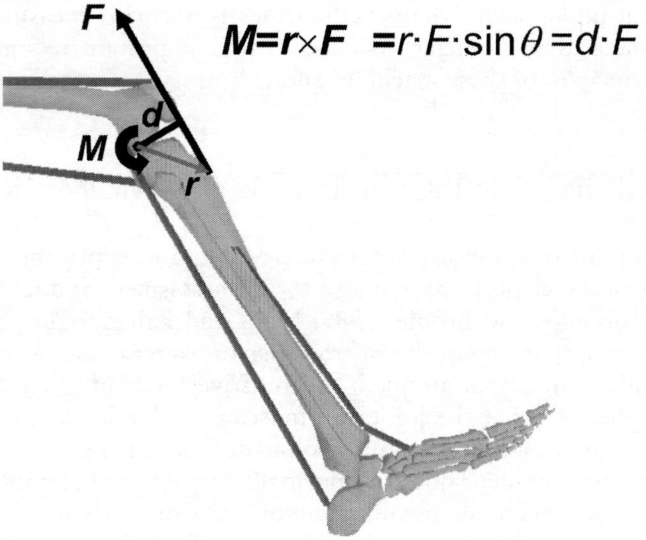

Figure 8.1 The application of a muscle force *F* (N) (knee extensor force transmitted via the patellar tendon in this example) around the axis of rotation with a position vector *r* relative to the origin. This generates a muscle moment *M* (N m) that is equal to the cross product (shown by the symbol ×) of the two vectors (*r* and *F*). The shortest distance between the force line of action and the axis of rotation is the moment arm *d* (m). *θ* is the angle between *r* and *F*. The moment *M* is also a vector that is perpendicular to the plane formed by *F* and *r* (coming out of the paper) and for this reason it is depicted by a circular arrow.

In general, the angular acceleration of a rotating segment will depend on the net moment applied, according to the fundamental equation of planar (two dimensional) motion for rotation:

$$\sum_{i=1}^{n} \mathrm{M_i} = \mathrm{I} \cdot \alpha \tag{8.1}$$

where:

$M_1 + M_2 + \cdots + M_n$: the total or net joint moment.
I: the moment of inertia (a parameter describing the distribution of the segment mass around the axis of rotation).
α: the angular acceleration of the rotating segment around the joint axis of rotation.

Muscle strength is the capacity to generate maximum force but since muscles generate joint rotation, strength is normally defined as the maximum joint moment that can be exerted under different muscle length, velocity and action (concentric, isometric, eccentric) conditions. For this reason, the measurement of net joint moment at different joint positions, angular velocities and muscle action conditions is essential for the assessment of muscle strength and the dynamic capabilities of the neuromuscular system in general. The maximum joint moment under such dynamic conditions is normally measured using different isokinetic dynamometer systems, so it is important to understand the operating principles of these machines and the measurement techniques used.

APPLICATIONS OF ISOKINETIC DYNAMOMETRY

Muscle and joint function assessment are essential in sport and exercise, not only for performance purposes but also for the assessment and rehabilitation of injury (Baltzopoulos and Brodie, 1989; Kellis and Baltzopoulos, 1995). Isokinetic dynamometry is one of the safest forms of exercise and testing. Once the pre-set angular velocity is attained, the resistive moment is equal to the net moment applied, so that the joint and muscles are loaded to their maximum capacity over the range of constant (isokinetic) movement. Because the dynamometer resistive moment does not normally exceed the net applied moment, no joint or muscle overloading and, therefore, risk of injury occurs.

Measurement of the joint moment, exerted at different angular velocities and joint positions, enables the establishment of moment velocity and moment position relationships for various joints in people from different populations. This information is essential for computer modelling and simulation of human movement (e.g. King and Yeadon, 2002; Yeadon et al., 2014) as well as for research into the physiology and mechanics of muscle (see Kellis and Baltzopoulos, 1996). Major review papers and books that describe isokinetic dynamometry applications in dynamic muscle strength assessment, rehabilitation and clinical problems include: Osternig (1986), Baltzopoulos and Brodie (1989), Cabri (1991), Davies (1992), Perrin (1993), Kannus (1994), Kellis and Baltzopoulos (1995), Chan et al. (1996), Gleeson and Mercer (1996), Brown (2000), De Ste Croix et al. (2003) and Dvir (2004).

This chapter will discuss mainly the biomechanical aspects of isokinetic movements and measurements; for consideration of physiological aspects of isokinetic dynamometry, such as muscular adaptations with training and relationships with other physiological measurements, see for example Brown (2000).

MECHANICAL BASIS OF ISOKINETIC DYNAMOMETRY MEASUREMENTS

An isokinetic dynamometer is a rotational device with a fixed axis of rotation that allows rotation with a constant angular velocity that is user selected. Figure 8.2 is a simplified schematic diagram of the basic components and operation of a typical isokinetic dynamometer. The user applies a force on the input arm as shown in Figure 8.2. This causes its rotation because it creates a moment, which is the product of the force applied on the input arm by the limb segment and its moment arm (shortest distance from the axis of rotation). The instantaneous angular velocity is monitored continuously and the braking mechanism is engaged accordingly, so that the required level of angular velocity is kept constant. The input arm is allowed to accelerate or decelerate if the instantaneous

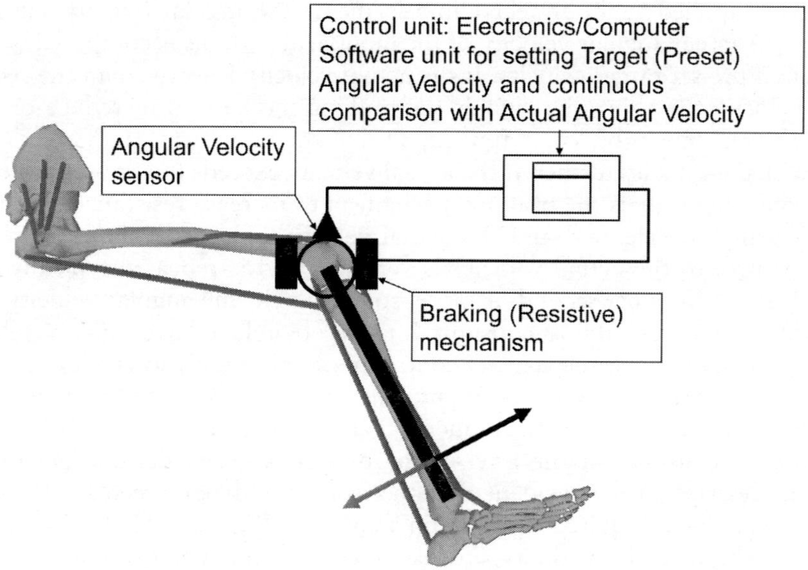

Figure 8.2 Schematic simplified diagram of the main components of an isokinetic dynamometer.

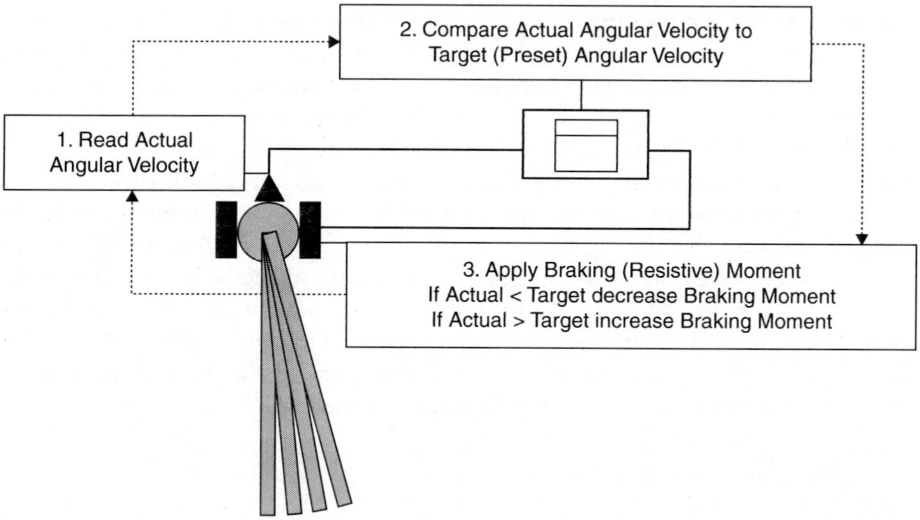

Figure 8.3 Schematic simplified diagram of the feedback loop for the control of the angular velocity by adjusting the resistive moment applied by the braking mechanism of the dynamometer. The resistive moment exerted against the limb depends on whether the actual angular velocity of the input arm is higher or lower compared to the user selected target (pre-set) angular velocity.

velocity is lower or higher than the pre-set target angular velocity, respectively. This represents a closed feedback loop mechanism with the actual (instantaneous) angular velocity as the controlled variable. Figure 8.3 describes the main steps in a typical feedback control loop in isokinetic machines.

The user selects the angular velocity required for the test (pre-set target velocity). Once the input arm starts rotating under the influence of an external moment (applied by the attached limb segment), the angular velocity sensor registers the actual angular velocity of the input arm and this is compared with the required pre-set target velocity. If the actual velocity is lower than the required test velocity the controller signals the braking mechanism to reduce or maintain the resistance (braking) applied, so that the input arm can accelerate and increase the velocity further. If the actual velocity exceeds the pre-set target then the controller signals the braking mechanism to increase resistance, so that the input arm is decelerated and the actual velocity reduced. This feedback loop and control of the actual velocity is very fast and is repeated typically about 1000–2000 times per second or more, so effectively, the angular velocity of the rotating arm is kept almost constant. It is not possible to have a completely constant velocity because the velocity control mechanism is always trying 'catch-up' with the actual velocity of the input arm. However, because the control loop is repeated with such a high frequency (>1000 Hz), the actual velocity is kept practically constant, at the level of the pre-set target velocity. From the preceding description it is evident that the isokinetic dynamometer provides the resistive moment required to maintain a constant angular velocity. The resistive mechanism depends on the type of isokinetic dynamometer; it could be based on hydraulics, electromechanical (motors) or a combination of the two types of components.

The resistive moment exerted by the braking mechanism, or the equivalent moment applied on the input arm by the segment, is provided as the main dynamometer system output together with the angular position of the input arm, usually via a dedicated computer system or analogue display devices on the dynamometer. However, what is required for the quantification of dynamic muscle strength is the actual joint moment applied by the muscles around the joint axis of rotation. It is very important to understand how this isokinetic dynamometer resistive moment output is related to the joint moment applied by the user because this has significant implications for the estimation of the joint moment and muscle strength and the interpretation of the isokinetic dynamometer data in general. The angular motion of the input arm around the fixed axis of rotation is governed by the same equation of planar motion for rotation (see equation 8.1 previously), so in a free body diagram of the dynamometer input arm for a knee extension test (Figure 8.4, left panel):

$$M_S - M_D - M_{GD} = I_D \cdot \alpha_D$$

where:

M_S: The moment of the segment force F_S around the axis of rotation of the dynamometer
M_D: The resistive moment exerted by the dynamometer
M_{GD}: The moment of the gravitational force F_{GD}
I_D: The moment of inertia of the dynamometer input arm
α_D: The angular acceleration of the dynamometer input arm

Assuming that the gravitational moment M_{GD} is measured and that the rotation is performed at an approximately constant velocity ($\alpha_D \approx 0$), then based

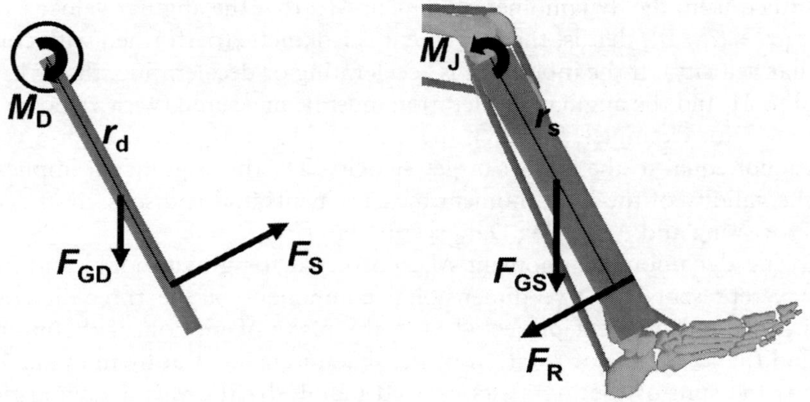

Figure 8.4 Free body diagrams of the dynamometer input arm (left) and the segment (right) for a knee extension test. Muscle strength is assessed by estimating the joint moment M_J from the dynamometer-measured moment M_D.

on the preceding equation M_S can be calculated from M_D. So by monitoring M_D from the dynamometer during the constant velocity period we have effectively a measure of the M_S moment applied by the segment on the input arm and causing its rotation.

The angular motion of the segment around the joint axis of rotation is governed by the same equation of planar motion for rotation (see equation 8.1) so in a free body diagram of the segment (Figure 8.4, right panel):

$$M_J - M_R - M_{GS} = I_S \cdot \alpha_S$$

where:

M_J : The joint moment exerted mainly by the active muscle groups spanning the joint
M_R : The moment of the dynamometer resistive force F_R around the axis of rotation of the joint
M_{GS} : The moment of the gravitational force F_{GS}
I_S : The moment of inertia of the limb segment
α_S : The angular acceleration of the limb segment

Assuming that M_{GS} is also measured and accounted for, and that the rotation is performed at an approximately constant velocity ($\alpha_S \approx 0$), then M_J can be calculated from M_R. Given that M_R is equal and opposite to M_S (since F_S and F_R are action and reaction and $r_d = r_s$) and M_S can be derived from M_D, we can then quantify M_J from M_D given the preceding gravity moment compensation and constant velocity assumptions. Surprisingly perhaps, the angular velocity, although measured and controlled, is not usually an output variable in most of the standard dynamometer software systems. This creates a number of important practical problems that affect the validity of measurements. As discussed earlier, there is a very important prerequisite for the joint moment M_J to be quantified from the dynamometer moment M_D: that the angular velocity is constant ($\alpha_D = \alpha_S = 0$), that is, the movement is isokinetic (performed with constant angular velocity). If the input arm is accelerating or decelerating, then M_D is not equal to M_J and the angular acceleration must be measured (with video or accelerometers for example), but even then, the adjusted M_D could be at a velocity that is not equal to the pre-set target velocity. This has significant implications for the validity of the joint moment measurements and is discussed in detail in the Processing and Analysing Data section later.

The dynamometer moment M_D, corrected for gravitational and inertial effects, represents the two-dimensional component of the three-dimensional joint force and moment projected onto the plane of motion of the input arm around the fixed axis of rotation of the dynamometer (Kaufman et al., 1995). This is the sum of the moments exerted mainly by the various agonistic and antagonistic muscle groups and the ligaments around the different axes of rotation of the joint (e.g. extension–flexion, internal–external rotation, adduction–abduction). Although the dynamometer is designed to assess isolated joint motion around one of these axes during a particular test, these joint axes are affected by the moments at the joint. They also do not coincide with the fixed axis of rotation of the dynamometer throughout the range of motion, even after careful positioning and stabilisation of the participant. Movements around these axes are also interrelated. These joint-dynamometer axes misalignment problems will be discussed in detail later. It is also evident from the preceding that the recorded dynamometer moment is not to be confused with the moment of the muscle groups or individual muscles. Approximate estimation of the many individual muscle and ligament forces acting around joints during isokinetic testing is only possible using appropriate reduction or optimisation techniques for the distribution of the recorded moment to the individual force producing structures (Kaufman et al., 1991). Some important problems and limitations with these techniques are discussed in relevant reviews (e.g. Tsirakos et al., 1997). It follows that moment–velocity or moment–position relationships from the dynamometer measurements refer to the joint in general. They must not be confused with the force–velocity or force–length relationships of isolated muscles or a muscle group, even if that group is the most dominant and active during the movement.

Torque and moment definitions

Torque and moment (or moment of force more precisely) are both terms that are used to express the rotational effect of forces applied around an axis of rotation. They are normally used interchangeably in physics and in the isokinetic dynamometry literature; this quite often creates some confusion amongst readers and users of isokinetics. Mechanically they usually describe different force application situations relative to the long structural axis of a segment (see Figures 8.5 and 8.6). Both terms are appropriate depending on whether we consider the structural effects of the force on the input arm (bending moment) or the twisting effect on the central rod of the dynamometer (twisting moment or torque).

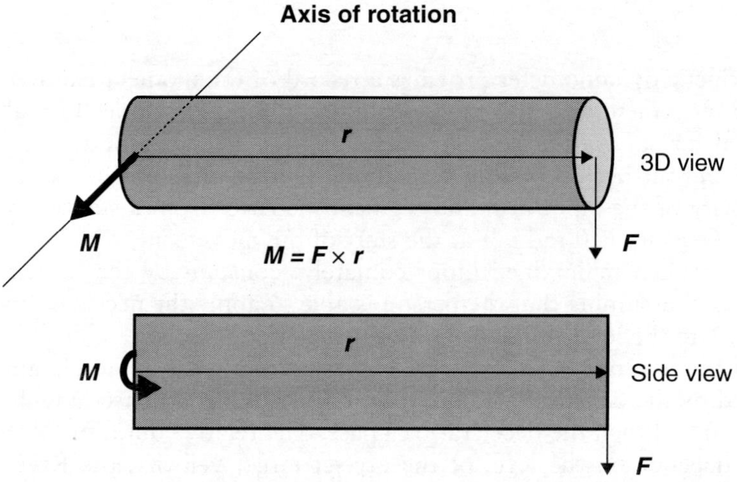

Figure 8.5 The definition of a moment (bending moment). Force vector and moment are perpendicular to the long structural axis.

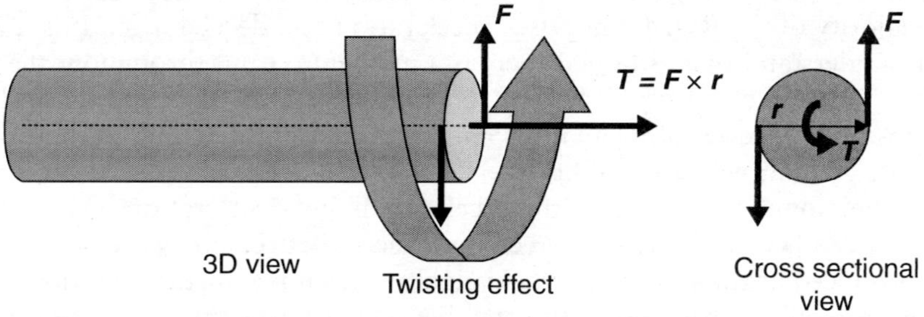

Figure 8.6 The definition of a torque (twisting moment) and the twisting effect. The axis of rotation is aligned with the long structural axis and the force pair is causing the torque. The torque vector is in line with the long structural axis and the axis of rotation.

Because we are concerned with the moment applied on the input arm by the attached segment and the joint moment, the more appropriate term for this purpose and the one that will be used throughout this chapter is moment (measured in N m or normalised to body mass as N m kg^{-1}). Reference to torque will be made only when specifically referring to the twisting of the dynamometer central rod.

Main factors affecting isokinetic measurements

There are a number of factors affecting the quality and the validity of isokinetic measurements, so it is important to consider the following when using isokinetic dynamometers.

Angular velocity

An isokinetic dynamometer provides a record of the moment exerted throughout the movement and this is presented to the user on line (in real-time) or in digital format via the dedicated data analysis and presentation software, if storage and further processing is required. During an isokinetic test, the angular velocity of the input arm and segment increases from a stationary position (angular velocity is 0 rad s^{-1}) at the start of the movement to the target pre-set value, it is then maintained (approximately) constant by the velocity control mechanism, assuming that the person is able to apply the necessary moment to accelerate to the level of the pre-set velocity, and then decreases back to zero at the end of the range of motion (ROM). There is always an acceleration phase, followed by the constant angular velocity or isokinetic phase (angular acceleration ≈ 0) and then the deceleration phase. The relative duration of these three phases depends on the level of the pre-set target velocity, the ROM and the capabilities of the person and any pre-activation of the muscles before the start of the movement. A number of studies have shown that the duration of the constant velocity (isokinetic) phase is reduced with increasing angular velocity tests, as expected. During very high angular velocity tests, the constant velocity or isokinetic phase is very limited (or non-existent in very weak participants) and the majority of the ROM comprises acceleration and deceleration. It is important to understand that although there is a moment output throughout the three phases, only the moment during the constant velocity (isokinetic) phase should be considered and used for analysis and assessment of dynamic strength at the required angular velocity conditions.

The moment data during the acceleration and deceleration phases should be discarded because, even if corrected for inertial effects, they are not recorded and produced at the required target velocity and in isokinetic conditions. This, however, requires checking of the angular velocity over the same timeframe as the moment output so that the instantaneous angular velocity for each moment value is known. Figure 8.7 is such an example of moment–time and angular velocity–time displayed together to allow a simple check of the moment data portion that was recorded under isokinetic (constant angular velocity) isokinetic (constant angular velocity) conditions. Despite the importance of this check, the moment and angular velocity are not very often displayed together in the standard dedicated computer software of most isokinetic machines, even though the angular velocity is monitored and recorded in all isokinetic systems. If this option is not available in the dedicated computer software then the user must perform this check independently. This can be done in a number of ways depending on the dynamometer and software used. For example, the CYBEX Norm model has an auxiliary interface card that can be installed separately and it allows access to the analogue data output of torque, angular position, angular velocity and direction of movement. The user can sample these analogue signals using a standard analogue to digital (A/D) conversion system and display them together in a separate computer system used for data collection. The dedicated computer-software system is still required for dynamometer set-up and control. In more recent software updates for the CYBEX range of dynamometers (HUMAC

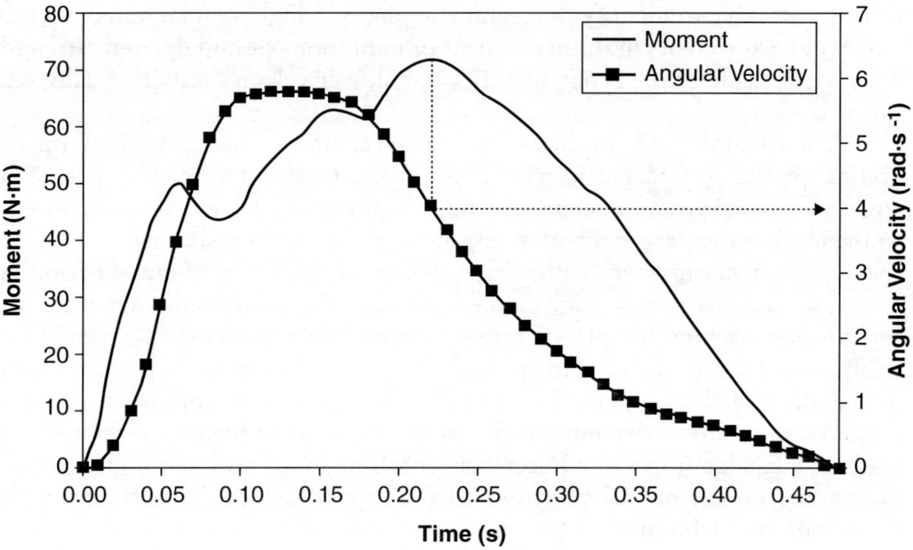

Figure 8.7 Moment and angular velocity during a knee extension test with the pre-set target velocity set at 5.23 rad s^{-1} (300 deg s^{-1}). Notice that the maximum moment was recorded when the angular velocity was just under 4 rad s^{-1} during the deceleration (non-isokinetic) period.

system) there are enhanced options for exporting all data after collection, including angular velocity. In other systems, such as the BIODEX dynamometer, the angular velocity signal is sampled and is also available for exporting via the standard software. In more recent versions of the Biodex software (System 4), the constant velocity or isokinetic periods are identified automatically and the user has the option of 'windowing' the data and considering only the isokinetic phase of the movement. Further details of the sampling resolution and accuracy characteristics of these hardware and software components are provided in the following section, Isokinetic Equipment Considerations.

The angular velocity values will contain some noise (from the A/D conversion and/or the differentiation of the angular position data if the angular velocity is not recorded directly). Furthermore, as explained earlier, the velocity control mechanism will always have some very small time delay ('trailing behind') so it is not possible to have a numerically exact constant value even in the constant angular velocity (isokinetic) period. For these reasons the angular velocity data require some filtering (see Chapter 9) and the isokinetic phase must then be determined with a reasonable limit for the filtered angular velocity fluctuation. This is normally between 5 and 10 per cent of the target pre-set value to accommodate different velocity control mechanisms and noise levels.

If an isokinetic dynamometer is used for research or any kind of dynamic strength assessment, especially when high joint angular velocities are used (>3–4 rad s^{-1}), then it is essential to have access to the digital or analogue angular velocity signal through the standard or any available auxiliary output. Simultaneous checking of the moment and the corresponding angular velocity over time is necessary to identify the isokinetic phase in high angular velocities tests and avoid gross errors in the assessment of joint moment and dynamic strength. This is not required, or is not as critical, at slower velocity tests (typically <2.1 rad s^{-1}).

The operation of the dynamometer mechanism during the initial acceleration period in concentric tests is another important mechanical issue. In some isokinetic dynamometers, the body segment is allowed to accelerate up to the level of the pre-set angular velocity without much resistance. In this case the dynamometer moment will be very close to zero for a brief initial period but obviously there will be a joint moment accelerating the system. When the segment is approaching the pre-set target velocity, the resistive mechanism is suddenly activated and required to apply a large resistive moment to prevent further acceleration of the segment. This initial resistive moment appears as a prominent overshoot in the dynamometer moment recording ('torque overshoot') and leads to a sudden impact and deceleration of the input arm and body segment. Depending on the pre-set velocity and the capabilities of the participant, this overshoot could be much higher than the maximum moment of the joint at that position. Modern dynamometers apply some resistive moment during the acceleration phase, irrespective of the actual angular velocity magnitude, and this helps to avoid the need for a sudden application of a large resistive moment as the segment approaches the pre-set velocity target. This response of the braking mechanism and the amount of resistive moment applied at the beginning of the movement, until the pre-set velocity is reached, is normally adjustable and is described as 'cushioning'. See the section on Processing and Analysing Data for the effects of these options on the isokinetic parameters. It is important to note that the preceding description of the angular velocity control is relevant for the concentric mode of dynamometer operation since in the eccentric mode, the speed of the motor is controlled independently of any moment applied by the participant.

Gravitational moment

The total moment applied around the dynamometer axis and recorded by the torque transducer is the result of the effect of all the moments applied on the input arm, including the gravitational moments due to the weight of the input arm itself and the attached segment. For example, if the input arm is locked (for isometric testing) in the horizontal position with the attached limb segment relaxed, there will still be a flexion moment recorded due to the moments caused by the weights of the input arm and the limb segment. The total gravitational moment will be a function of the cosine of the input arm angle, being maximum in the horizontal position (0°) and zero in the vertical position (90°) – see Figure 8.8.

Most isokinetic systems include an automated gravity moment correction procedure as part of the dedicated software system. This involves the measurement of the total gravitational moment during a passive weighing of the segment either throughout the range of motion or at a mid-range position. The moment recorded during the test is then adjusted by either subtracting or adding

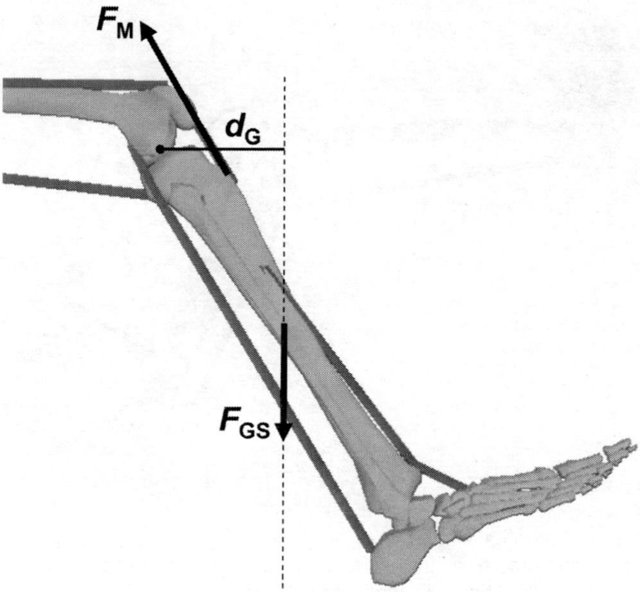

Figure 8.8 Gravitational moment due to the weight of the segment (F_{GS}) acting with a moment arm d_G around the axis of rotation of the joint. Since the gravitational force is constant, the gravitational moment will depend on d_G and will be maximum at full extension and zero with the segment in the vertical position (90° of knee flexion in this example).

(depending on the direction of the input arm movement) the measured or calculated moment due to gravity at the different positions.

It is important to point out that the gravitational effect is an issue only for movements in the vertical plane. If it is technically possible for the dynamometer main unit to be tilted so that its axis of rotation was vertical and the segment plane of motion was horizontal (e.g. elbow flexion – extension with the lower arm attached to the input arm of the dynamometer and the upper arm abducted 90° and horizontal) then the gravitational moments would not affect the moments recorded by the dynamometer.

Axes of rotation alignment

Isokinetic dynamometers are designed to measure the moment applied around their fixed axis of rotation. Therefore, to exert the maximum rotational effect (moment) possible, the force applied must be perpendicular to the long axis of the input arm. If an external force is applied by the segment at an angle other than 90° to the input arm then only the perpendicular component will be generating a moment (Figure 8.9), since any component of the applied force along the input arm long axis will be acting through the axis of rotation of the dynamometer so no moment will be generated (moment arm = 0).

It was shown earlier in this section, using the free body diagrams of the segment and the input arm, that one of the necessary conditions for using M_D from the dynamometer to quantify the joint moment M_J is that $M_S = M_R$ (see

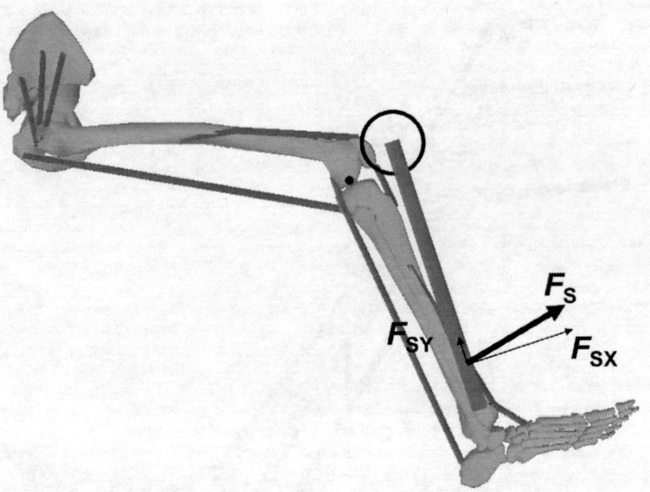

Figure 8.9 Effects of misalignment of axes of rotation. The axes of rotation of the segment and dynamometer input arm are not aligned and, in this case, the long axes of the segment and input arm are not parallel either. Because the segment attachment pad rotates freely and is rigidly attached to the segment, the force applied by the segment (F_S) is perpendicular to its long axis but not perpendicular to the dynamometer input arm. As a result, only a component (F_{SX}) of the applied force F_S is producing a moment around the axis of rotation of the dynamometer.

Figure 8.4). Since by definition F_R is equal and opposite to F_S (action–reaction) then the moment arms of these two forces (r_s and r_d) must also be equal.

The only way for $r_s = r_d$ is for the axes of rotation of the segment and the dynamometer to be aligned. If the joint axis of rotation is not in alignment with the dynamometer axis then r_s will be different to r_d and therefore $M_S \neq M_R \Rightarrow M_J \neq M_D$ so we will not be able to quantify the joint moment from the dynamometer moment.

For example, let us assume that the long axes of the segment and the input arm overlap on the sagittal plane, but the joint axis of rotation is located 2 cm above the dynamometer axis of rotation and the input arm is attached 30 cm from the joint axis so that $r_s = 0.3$ m but $r_d = 0.28$ m (Figure 8.10).

If the joint moment exerted was known and was $M_J = 270$ N m then the force F_S applied to the input arm and its reaction applied on the segment F_R will be $F_S = F_R = 900$ N: $M_J = M_R = 900 \times 0.3 = 270$ N m.

The 900 N applied on the input arm of the dynamometer, 28 cm from its fixed axis of rotation will be generating a moment $M_S = 900 \times 0.28 = 252$ N m and this will be registered by the dynamometer. So this 2 cm error in the alignment of the axes of rotation will be causing an error of 18 N m (6.6 per cent) in the measurement of joint moments: ((270–252) / 270) × 100.

In general if $r_d \neq r_s$ then (ignoring gravitational and inertial forces)

$$M_J = F_R\, r_s \text{ and } F_R = M_J / r_s$$
$$M_D = F_S\, r_d \text{ and } F_S = M_D / r_d$$

and since $F_R = F_S$ then $M_J / r_s = M_D / r_d$ and

$$M_J = M_D\, (r_s / r_d)$$

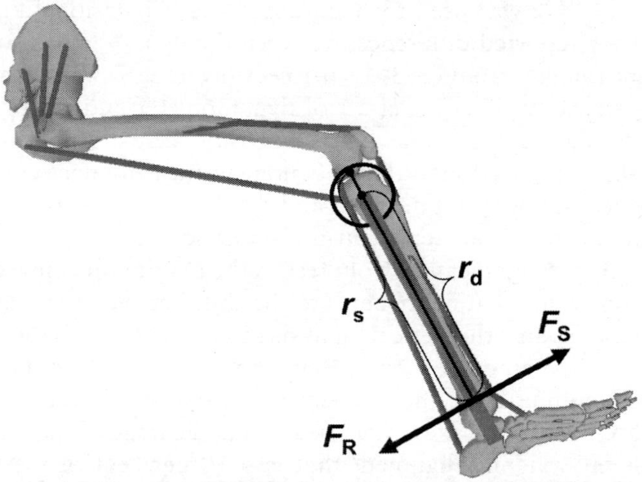

Figure 8.10 An example of dynamometer and joint axis of rotation misalignment. In this case, the long axes of the segment and input arm are parallel (coincide in 2D) so the force applied by the segment F_S is perpendicular to the input arm but the moment arms of the forces F_S and F_R relative to the dynamometer (r_d = 0.28 m) and joint (r_s = 0.3 m) axis of rotation, respectively, are different. As a result, the joint moment (M_J) and the dynamometer-recorded moment (M_D) are also different. (Gravitational forces are ignored in this example).

In general then, if the moment arms of the segment and the dynamometer are different because of the misalignment of joint and dynamometer axes of rotation, then the dynamometer moment M_D must be corrected by multiplying it with the ratio of the moment arms (r_s/r_d). This is the only way to measure the joint moment accurately. Although the moment arm of the input arm force does not change (r_d is a constant), the moment arm of the segment will be changing based on the position of the joint centre (axis of rotation) throughout the motion. In the knee, for example, a number of studies have examined the error caused by the misalignment of the axes of rotation of the joint and the dynamometer during contraction. Herzog (1988) examined a single participant using a Cybex II dynamometer and a two-dimensional kinematic analysis system filming at a frequency of 50 Hz. He reported that the maximum differences between the actual knee joint moment and the recorded moment during isometric and isokinetic knee extension trials at 120 and 240 deg s^{-1} were 1.5, 1.3 and 2.1 per cent, respectively. Kaufman et al. (1995) quantified the errors associated with the misalignment of the knee joint axis and the CyBex II axis of rotation during isokinetic knee extensions on five normal participants. They measured the three-dimensional kinematics of the leg using a tri-axial electrogoniometer and found that the average differences between the actual and recorded knee joint moments were 10 and 13 per cent during isokinetic knee extensions at 60 and 180 deg s^{-1}, respectively.

More recently, Arampatzis et al. (2004) determined the differences between the actual moment and the moment measured by dynamometry at the knee joint on 27 athletes during isometric knee extension. A Biodex dynamometer was used for the quantification of joint moments and a Vicon system with eight cameras operating at 120 Hz was used for recording three-dimensional kinematic data of the leg. They reported differences between the dynamometer and the actual joint moment ranging from 0.33–17 per cent (average 7.3 per cent). In a more recent study that examined the effect of dynamometer and joint axis misalignment using X-ray video on measured isometric knee-extension moments (Tsaopoulos et al., 2011) we found that the moment measurement error (difference between the actual knee joint moment and the dynamometer recorded moment) ranged from 1.9 to 4.3 per cent. During isometric knee extension, the internal knee angle changed significantly from rest to the maximum contraction state by about 19°, something that contributes to the shift between the segment and the dynamometer axis. In a theoretical analysis of the error due to the misalignment of the joint and dynamometer axes, Reimann et al. (1997) found that, in general, the maximum error in the moment measurement is roughly proportional to the displacement of the axes expressed as a percentage of the segment length. For example, an axis misalignment that is ~ 10 per cent of the length of the segment leads to approximately a 10 per cent maximum error in the moment measurement. The actual error with a 10 per cent shift in the axes ranged from –9 to 6 per cent in the range of motion from 0° (full extension) to 100° of knee flexion.

It must be emphasised that the positioning and stabilisation of the participants in all of the preceding studies were performed following the manufacturer's recommendations, using the standard belts, straps and other mechanical restraints supplied. Furthermore, the reported joint moment errors were totally inconsistent between participants and testing sessions and depended on the stabilisation of the participant, the movement of the isolated and moving segments relative to the dynamometer due to the compliance of the soft tissue and the dynamometer components, such as the seat and input arm padding. Other factors contributing to the random nature of the errors present in the joint moment measurement are, the initial alignment of the axes, the mode and intensity of contraction and the range of motion.

It is important to understand that there will always be a misalignment of the axes because: (1) joints do not have fixed axes of rotation and the changing instantaneous joint axis cannot be aligned with a fixed axis (dynamometer) throughout the ROM, and (2) no matter how carefully the joint axis is aligned with the dynamometer axis in a specific position, there will always be some movement of the segment relative to the input arm during the movement. Therefore, there will always be some change of the segment moment arm due to the compliance and compression of soft tissues (muscles, subcutaneous fat, etc.) and the dynamometer components (chair and input arm attachment padding). The wide range of the joint moment errors reported in several studies (e.g. 0.33–17 per cent by Arampatzis et al., 2004) within the same group of participants tested by the same investigators using standardised and recommended procedures for a relatively simple and highly constrained contraction mode (isometric) shows the scale of the problem, and the important implications if this error is not minimised during the data collection procedures, or corrected afterwards. The recommended techniques to deal with these problems and minimise the misalignment error are discussed later in the participant positioning section.

Adjacent joint position and bi-articular muscle length effects

Isokinetic dynamometry involves single joint testing so only the tested joint is allowed to move over its range of motion. All the other joints should be fixed and the segments stabilised to avoid extraneous movement and possible contributions. However, when bi-articular muscles are involved and these span the tested joint, it is important to consider carefully, record and standardise the position of the adjacent joints spanned by the bi-articular muscles. This is because the muscle force and, as a result, its contribution to the joint moment depends on the length of the muscles according to the force–length relationship as well as the velocity and mode of action as explained previously. For example, hip angle must be standardised during knee extension to avoid variations in the length of the rectus femoris, as it has been shown that the knee extension moment is different between supine and seated positions (e.g. Pavol and Grabiner, 2000; Maffiuletti and Lepers, 2003). Similarly, during a knee flexion test on the dynamometer, the length of the hamstrings will also be affected by the hip angle and chair position. Furthermore, the angle of the ankle joint (plantarflexion or dorsiflexion) during a knee flexion test will affect the length of the bi-articular gastrocnemius, which is a muscle that can also contribute to knee flexion moment under sufficient length conditions. It has been shown, for example, that knee flexion moment is higher with the knee flexion test performed with the ankle in dorsi-flexion (longer gastrocnemius length) compared to knee flexion with the ankle in full plantarflexion (Croce et al., 2000). So in this case not only the angle of the hip joint, but also the ankle joint angle, should be standardised and controlled.

ISOKINETIC EQUIPMENT CONSIDERATIONS

This section describes the main considerations when choosing an isokinetic device. The isokinetic equipment market has been quite volatile in recent years because of the availability of an increasing number of different machines from established and new manufacturers and the saturated market for isokinetic equipment in rehabilitation centres, universities and other clinical research facilities. As a result, a number of manufacturers have ceased trading altogether or they have closed down the isokinetic equipment part of the business and no further support or products are offered. This chapter will only consider isokinetic equipment, which, at the time of writing, is still available from authorised companies that can offer service and maintenance or repair support. A useful web resource that explains in detail the history of isokinetic equipment development from different manufacturers and also includes other useful information on using isokinetic equipment can be found at http://isokinetics.net/.

Isokinetic hardware and software

All the modern available isokinetic dynamometers use electromechanical components for the control of the angular velocity of the input arm. Hydraulic based systems are no longer available, with the exception of used KinCom models that are based on a combination of electromechanical and hydraulic components. Older models (e.g. Cybex II) had a passive resistive mechanism and resistance was developed only as a reaction to the applied joint moment. This mechanism allowed only concentric muscle action. The modern electromechanical dynamometers (e.g. Biodex, CON-TREX, Cybex NORM) have active mechanisms as well as the standard passive operation in the concentric mode. These dynamometers are capable of driving the input arm at a pre-set angular velocity irrespective of the moment applied by the person being tested. This allows eccentric muscle action at constant angular velocities. The operation of the resistive mechanism, and the control of the angular velocity and acceleration in these systems, requires the use of appropriate dedicated computer systems. However, there are considerable differences in the techniques implemented in the software of various dynamometers. These differences concern the collection and processing of moment data and the calculation of different mechanical variables, such as the gravitational moment, joint moment, angular velocity of motion, and moment developed in the initial acceleration period. For these reasons, the software-computer system features, user friendliness and any options must all be examined in detail to ensure that they are the best for the intended use of the equipment (research, rehabilitation, teaching, or combination of uses). Some dynamometers offer control via both the computer system and an analogue or digital control panel (e.g. older Biodex systems) and this can be very useful, for example, in teaching/demonstrating applications.

Angular velocities range and moment limits

The range of angular velocities available under concentric and eccentric modes and the respective maximum moment limits are some of the most important features in an isokinetic dynamometer. Table 8.1 summarises these characteristics in some of the most popular commercially available isokinetic dynamometers.

Isokinetic data resolution and accuracy considerations

The quality, reliability and hysterisis characteristics of the modern transducers used in the new generation of isokinetic dynamometers are very good. The accuracy of the torque, angle and angular velocity data has improved considerably in modern systems and typical values in different machines range from ±0.25 to ±1 per cent of full scale for torque and less than about 0.3 per cent of full scale for angle or from less than ±0.1° to ±1°. One of the most important considerations is access to the analogue or digital data of angle, torque and angular velocity, especially when the isokinetic dynamometer is used for research purposes. Older systems provided digital data with limited sampling rates (100–120 Hz) and access to the analogue data was only possible with optional accessory components at an extra cost. For example, as stated previously, it is possible to

Table 8.1 Summary of the range or limits of angular velocities and moments under concentric and eccentric modes for the most popular commercially available isokinetic dynamometers, including manufacturer website information.

Dynamometer	*Angular velocity (rad s⁻¹)*		*Moment limits (N m)*	
	C	*E*	*C*	*E*
Biodex www.biodex.com	0.5–8	0.1–2.6	610	410
CON-TREX www. con-trex.com	8.7	8.7	720	720
Cybex Norm www.csmisolutions.com	0.1–8.7	0.1–5.2	680	680
BTE PrimusRS www.btetech.com/		5.2–		226–
KinCom 500H www.kincom.com	0.02–4.4	0.02–4.4	800*	800*

Notes
C: concentric; E: eccentric. *Values obtained by assuming a moment arm of 0.

fit an auxiliary interface board to the later Cybex machines (e.g. NORM) that provides access to the analogue signals of torque, angular position, velocity and direction of movement. The analogue output can then be sampled using an A/D converter completely independently of the dedicated software/computer with a higher, user-determined sampling frequency. This is essential for research applications. Modern isokinetic systems (e.g. Biodex System 4, CYBEX NORM, CON-TREX) include 12- or 16-bit A/D converters resulting in a resolution of ~ ± 0.0244 per cent of full scale. Very high sampling rates of up to 4000 Hz or 5000 Hz further means that the resolution and accuracy of the digital data provided by the standard computer systems in these modern machines are quite adequate for research purposes.

Isokinetic equipment rigidity and space considerations

The overall strength and rigidity of the dynamometer and the stiffness of all the components are also very important considerations, especially when testing very strong athletes such as sprinters, rugby players, power lifters, etc. For example, it is unfortunately very common in older machines for the main unit to twist from its base where it is attached on the frame of the dynamometer when very strong athletes apply high moments, resulting in an unsafe situation and certainly invalid measurements. Modern dynamometers are heavier and more rigid with stronger and better supporting frames for the chair and main unit, but, ideally, the user must try a new machine with a sample of the likely population that they will be using in their work. The overall mass of isokinetic dynamometers is normally in the range of 500–700 kg and they typically require a minimum space of 6–7 m^2 for standard operation.

ISOKINETIC EXPERIMENTAL AND DATA COLLECTION PROCEDURES

The following factors can affect the overall quality of isokinetic data and have significant implications for assessment of joint function. They need to be considered in detail when setting up an isokinetic device for data collection.

Calibration

Moment calibration can be performed using gravitational loading under static or slow velocity conditions. Most manufacturers provide accurate calibration weights; to calibrate a large range of moments, a range of weights should be used. Accurate goniometers can be used to calibrate angular position. Several dynamometers also include a self-calibration mechanism for checking and adjusting the zero baseline or factory calibration of the torque transducer. Full calibration of the measurement system (moment and angle transducers) should be performed at the manufacturer recommended intervals or periodically, depending on usage, and recorded on a technical calibration log. Checking of the complete measurement and recording system (including interface boards, any amplifiers and analogue-to-digital converter) must be performed before each measurement session irrespective of the type of dynamometer used.

Familiarisation instructions

Constant angular velocity movements are rarely performed during exercise or sports activities; it is therefore important to familiarise the participants with this mode of joint action before the test. The complete range of the angular velocities to be used in the test must be included in the familiarisation session. Standardised instructions should be given, explaining that isokinetic testing involves variable resistance at a constant velocity and stressing the importance of maximum muscular effort during the tests. The maintenance of maximum effort, throughout the range of movement, by both muscle groups during reciprocal movements, such as knee extension–flexion, must be specifically emphasised. Failure to include familiarisation with this mode of joint function may have serious effects on the reliability of measurements.

Participant positioning

The person being tested should be stabilised to ensure that the recorded moment is generated by the examined joint muscles only, without any contribution from other actions. During a typical single segment movement, such as knee extension–flexion, the opposite leg, the waist, torso and the arms should be stabilised with appropriate belts or harnesses, usually provided by the manufacturers of the dynamometer. The examined body segment should be securely fastened to the input arm of the dynamometer to avoid moment overshoots and impacts between the limb and the input arm. If bi-articular muscles are involved then the angles of the adjacent joints should be considered carefully and must be standardised and controlled during the test to avoid bi-articular muscle contribution errors (see earlier discussion in the section 'Adjacent joint position and bi-articular muscle length effects'). For example, ankle and hip joint positions should be standardised and recorded in knee flexion tests. It is particularly important if a retest is performed to have a detailed record of all the test settings to ensure high repeatability and accurate conclusions.

The axis of rotation of the dynamometer must be aligned carefully with the joint axis to minimise axes misalignment errors. Although the instantaneous axis of rotation of a joint is difficult to establish visually and varies in dynamic conditions, an approximation using relevant anatomical landmarks is essential. For example, the most prominent point on the lateral epicondyle of the femur as palpated externally on the lateral surface of the knee joint or the most prominent point on the lateral malleolus for the ankle. A laser pointing device, or some other alignment tool, can be used to improve the alignment of the dynamometer axis with the anatomical landmark, or other point that the joint axis of rotation is assumed to pass through. Given that there is a large change in joint position and angle between rest and contraction (e.g. Arampatzis et al., 2004; 2005), it is necessary to align the axes during submaximal or maximal contraction and not with the segment at rest. In order to minimise further the axes misalignment error in specific parameters such as the maximum joint moment, the active alignment (under maximal or submaximal contraction conditions) should be performed near the joint position where the maximum joint moment is expected or at the joint angle where any angle-specific moment will be measured.

Range of motion

The range of movement must be set according to the physiological limits of the joint and activated muscle group and any inhibiting injuries; it must also be standardised and controlled. Standardisation of the range of movement also facilitates data analysis, for example integration of the moment with respect to angular displacement to calculate mechanical work. The angular velocity of the movement should be set according to the objectives of the test and the muscular capabilities of those being tested. During fast velocity testing (>3.1 rad s^{-1}) it must be ensured that the person can achieve the pre-set velocity within the range of joint movement. The duration of the isokinetic (constant velocity) phase of the movement decreases as the test speed is increased, as more time is required to reach higher pre-set velocities and to decelerate at the end of the movement. This is important, as the dynamometer moment in concentric conditions is not equal to the joint moment if the pre-set velocity is not attained and kept constant, as explained earlier.

Preloading

Preloading of the muscles, using electrical stimulation or manual resistance before the release of the isokinetic mechanism, can be used to facilitate the development of faster joint velocities over limited ranges of movement.

Gravity correction

Most computerised dynamometers include gravity correction as part of the experimental protocol; this correction must be performed before each test of joint function to minimise measurement errors. If a computerised procedure is not available, a simple method for gravity correction should be used. One method involves the recording of the gravitational moment generated by the weight of the segment and lever arm falling passively against the resistance of the dynamometer, at a specific angular position within the range of movement. The gravitational moment at different angular positions is then calculated as a function of the gravitational moment at the measurement position and joint angle. This procedure must be performed at the minimum angular velocity or under isometric conditions. The limb must be relaxed and the muscle groups involved, both agonists and antagonists, must not be close to their maximum length, to avoid unwanted contributions from the elastic components. In practice, it is suggested that several trials should be performed to ensure complete muscular relaxation during this measurement. It has been shown, however, that this method gives an overestimation of the actual gravitational moment of the segment, because of the elastic effects of the musculotendinous unit and incomplete relaxation (Kellis and Baltzopoulos, 1996). Estimation of the gravitational moment from anthropometric data has been found to be a more accurate gravity correction method; it is therefore suggested that this method may be a useful alternative to the passive fall/limb weighing method. The gravitational moment at different angular positions is then added to the dynamometer moment produced by muscle groups opposed by gravity, for example the quadriceps femoris in a knee extension–flexion movement in the sagittal plane (see Figure 8.4 and related equations). For muscle groups facilitated by gravity, such as the hamstrings, the gravitational moment is subtracted from the recorded moment.

Performance feedback

Visual feedback of joint moment data during isokinetic testing has a significant effect on the maximum moment. The magnitude of this effect depends on the angular velocity of movement. It is therefore important to provide visual feedback of the muscular performance during the test, particularly during maximum moment assessment at slow angular velocities. The real-time display of the moment output on a computer monitor, or other display device, can be used for visual feedback. It is important to give detailed instructions on the interpretation of the different sources of visual feedback and the performance target. Verbal motivation for maximum effort must be given in the form of standardised and consistent instructions/encouragement.

Control and standardisation of experimental procedures

Differences in the experimental factors described in this section can significantly affect moment measurements. Different isokinetic dynamometers also use different settings and testing procedures. It is evident from the preceding, that testing of the same person on different machines will most likely produce different results, because of biological and mechanical variations. Differences in data processing from different dynamometer software applications introduce further discrepancies into the assessment of joint moment. Joint function measurements are therefore specific to the dynamometer, the procedures and the data processing techniques used. Comparison of results from different dynamometers, even from the same person, may not be possible. For these reasons, it is essential to standardise and report (see Reporting an Isokinetic Study section) the relevant details of the experimental procedures and data processing; it is particularly important to use exactly the same settings in test–retest experiments.

In summary:

- Calibrate dynamometer frequently or check accuracy of data output.
- Familiarise participants and give clear and standardised, consistent instructions.
- Take particular care during eccentric tests and when testing children, or weak or injured/operated participants.
- Stabilise and control adjacent joints when bi-articular muscles are involved.
- Stabilise segments and reduce extraneous movement.
- Align axes of rotation:
    - as accurately as possible
    - under maximal or submaximal contraction conditions
    - near the position of expected maximum joint moment.
- Monitor angular velocity independently.

PROCESSING, ANALYSING AND PRESENTING ISOKINETIC DATA

Angular velocity monitoring and isokinetic data processing

To overcome measurement errors resulting from the overshoot in the moment recordings, analogue electrical filters have been used to dampen the overshoot and smooth the signal. Different cut-off frequencies (damping settings) have been suggested, depending on the angular velocity of the movement and the testing conditions. These electrical filters introduce a phase shift into the moment-time signal; the overshoot peak is reduced, and the filter also reduces the magnitude of the moment throughout the range of movement. The use of such filters must therefore be avoided, as it leads to significant distortion of the moment-angular position relationship that is the basis for the measurement of most isokinetic joint function parameters. Some isokinetic dynamometers (e.g. Biodex) apply a resistive moment from the start of the concentric movement, before the development of the pre-set velocity, just by sensing movement of the input arm. This reduces the acceleration rate and therefore the time required to attain the pre-set velocity is increased. However, the transition from acceleration to the pre-set constant angular velocity is smoother and no sudden resistive overshoot moment to decelerate the dynamometer input arm is required. The degree of acceleration control and the smoothness of the transition to the constant velocity (isokinetic) phase is adjustable in different dynamometers and this is usually determined by a 'cushioning' setting ranging from 'soft' (more acceleration control-smooth transition) to 'hard' (more sudden transition to the isokinetic phase).

Only the isokinetic part of the movement should be considered and moment data must be analysed from the constant angular velocity part of the movement only (Figure 8.11), irrespective of the method used to control acceleration and its effects on movement recording. Although several dynamometers

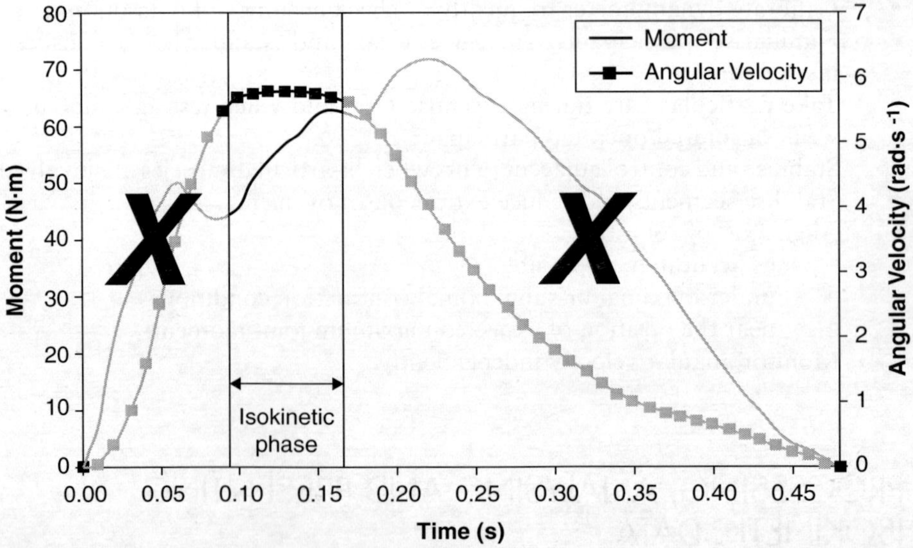

Figure 8.11 At high target velocities the isokinetic (constant angular velocity) movement is very limited or non-existent. In this test with the target velocity pre-set at 5.23 rad s^{-1} (300 deg s^{-1}), the isokinetic phase lasts only approximately 0.075 s, and is only about 15 per cent of the total extension movement. Moment data outside this interval should be discarded because they do not occur in isokinetic (constant angular velocity) conditions and the actual angular velocity of movement is always slower than the required pre-set velocity.

provide angular velocity data, it has been shown that these measurements may contain large errors, particularly during tests with a high pre-set angular velocity (Iossifidou and Baltzopoulos, 1998, 2000). It is therefore recommended that the user should independently calculate kinematic variables, such as the angular velocity and acceleration of the body segment, using the angular position time data provided by the dynamometer. The digital data contain errors (noise) mainly from the analogue-to-digital conversion; appropriate noise reduction techniques are therefore needed for the accurate measurement of kinematic variables. Several noise reduction and differentiation techniques (e.g. digital filters, spline functions and truncated Fourier series) that have been applied in other areas of biomechanics research (see Chapter 9) are appropriate.

Isokinetic parameters

Maximum moment

The maximum moment during an isokinetic movement is an indicator of the maximum muscular forces applied in dynamic conditions (dynamic strength). Various testing protocols have been used for the assessment of this maximum moment; it is usually evaluated from three to six maximal repetitions and defined as the maximum single moment value measured during these repetitions. The average value of the maximum moments from several repetitions should not be used as an indication of the maximum joint moment. The maximum moment depends on the joint angular position at which it was recorded, as this angle affects the force–length relationship of the active muscles. A maximum moment, calculated as the average of several maximum moment values recorded at different joint positions, is not an appropriate measure of muscle function, as it provides no information about the joint moment–angular position relationship. The average from several repetitions is only useful when the moment is recorded at a specific predetermined joint position in each repetition. In this case, however, the recorded moment at the predetermined position may not be the maximum moment for that repetition.

Reciprocal muscle group moment ratios

The reciprocal muscle group ratio is an indicator of joint balance; it is usually affected by age, sex and physical activity (Baltzopoulos and Brodie, 1989). The measurement of reciprocal muscle group ratios must be performed using gravity-corrected moment data to avoid errors in the assessment of muscle function.

Angle of maximum moment

The joint angular position is important in the assessment of muscle function because it provides information about the mechanical properties of the activated muscle groups. The maximum moment position is affected by the angular velocity of the movement; it tends to occur later in the range of movement with increasing velocity, and not at the mechanically optimal joint position. Furthermore, it is essential to check whether the maximum moment was recorded at a joint position in which the angular velocity was constant.

Fatigue and endurance

Muscular endurance under isokinetic conditions is the ability of the activated muscle groups to sustain a movement with maximum effort at the pre-set angular velocity. It is usually assessed by computing a fatigue index. However, there is neither a standardised testing protocol for the assessment of muscular endurance nor an agreed definition of the fatigue index. It has been suggested that muscular endurance should be assessed: over 30 to 50 repetitions; up to the point when the moment has fallen to 50 per cent of the maximum initial moment; or over a total duration of 30 to 60 s. The fatigue index can be expressed as a ratio between the maximum moment in the initial and final periods of the test. The mechanical work performed is a more representative measure of muscle function, because it is computed from the moment output throughout the range of movement. To compute the fatigue index from mechanical work measurements, however, the range of movement must be standardised and the moment data must be corrected for the effects of gravitational and inertial forces.

Acceleration related parameters

Some isokinetic parameters for the analysis of concentric muscle function during the initial part of the movement (such as joint moment development and time to maximum moment) have been based on dynamometer moment data during the acceleration period. There is a misunderstanding that the dynamometer moment during this initial acceleration period reflects the joint moment and, therefore, can provide a measure of the ability to develop muscle force rapidly. However, it is evident from the description of the isokinetic measurement process that, in concentric conditions, the dynamometer moment during this initial acceleration period represents the resistive moment developed by the dynamometer and not the actual joint moment accelerating the system. Furthermore, the resistive dynamometer moment in concentric conditions is delayed until the pre-set velocity is attained by the moving limb and depends also on the response of the resistive mechanism in different types of dynamometers and any relevant settings used (e.g. cushioning, damping etc.). Consequently, any conclusions about the mechanical properties of the muscle group, based on the dynamometer moment output during the initial acceleration period, are invalid. The joint moment during the acceleration and deceleration periods should not be used for any measurements (e.g. 'torque acceleration energy') and only the isokinetic phase moment should be analysed, as explained in the Angular Velocity Monitoring and Isokinetic Data Processing section.

REPORTING AN ISOKINETIC STUDY

Full details of the following should be included in the reporting of an isokinetic study:

- Isokinetic dynamometer make and type.
- Details of data acquisition: it is essential to report whether the standard software of the dynamometer or an independent system was used to sample the analogue signals. In this case the details of the sampling and A/D processes are necessary (sampling frequency, resolution and accuracy of moment and angular position measurements, etc.).
- Procedures followed for calibration and/or checking of measurements and gravity correction.
- Test settings (eccentric/concentric, range of movement, angular velocity, feedback type and method).
- Participant positioning: given the effects of axes misalignment, the method of axes alignment used must be described in detail. The angle of adjacent joints (e.g. hip and ankle joint position during knee tests) must also be reported.
- Isokinetic (constant angular velocity) phase determination: details of angular velocity monitoring method (dynamometer software or independent sampling) and determination of the constant velocity (isokinetic) phase. This can be omitted in slow velocity testing (<180 deg s^{-1}) but it is essential in high velocity testing.

REFERENCES

Arampatzis, A., Karamanidis, K., De Monte, G., Stafilidis, S., Morey-Klapsing, G. and Bruggemann, G. P. (2004) 'Differences between measured and resultant joint moments during voluntary and artificially elicited isometric knee extension contractions', *Clinical Biomechanics*, 19(3): 277–283.

Arampatzis, A., Morey-Klapsing, G., Karamanidis, K., DeMonte, G., Stafilidis, S. and Bruggemann, G. P. (2005) 'Differences between measured and resultant joint moments during isometric contractions at the ankle joint', *Journal of Biomechanics*, 38(4): 885–892.

Baltzopoulos, V. and Brodie, D. A. (1989) 'Isokinetic dynamometry: applications and limitations', *Sports Medicine*, 8(2): 101–116.

Brown, L. E. (2000) *Isokinetics in Human Performance*, Champaign, IL: Human Kinetics.

Cabri, J. M. (1991) 'Isokinetic strength aspects of human joints and muscles', *Critical Reviews in Biomedical Engineering*, 19(2–3): 231–259.

Chan, K. M., Maffulli, N., Korkia, P. and Li Raymond, C. T. (1996) *Principles and Practice of Isokinetics in Sports Medicine and Rehabilitation*, Hong Kong: Williams and Wilkins.

Conceição, F., King, M. A., Yeadon, M. R., Lewis, M. G. C. and Forrester, S. E. (2012) 'An isovelocity dynamometer method to determine monoarticular and biarticular muscle parameters', *Journal of Applied Biomechanics*, 28(6): 751–760.

Croce, R. V., Miller, J. P. and St Pierre, P. (2000) 'Effect of ankle position fixation on peak torque and electromyographic activity of the knee flexors and extensors', *Electromyography and Clinical Neurophysiology*, 40(6): 365–373.

Davies, G. J. (1992) '*A Compendium of Isokinetics in Clinical Usage and Rehabilitation Techniques*', 3rd edn. Onalaska, WI: S and S Publishers.

De Ste Croix, M., Deighan, M. and Armstrong, N. (2003) 'Assessment and interpretation of isokinetic muscle strength during growth and maturation', *Sports Medicine*, 33(10): 727–743.

Dvir, Z. (ed.) (2004) *Isokinetics: Muscle Testing, Interpretation and Clinical Applications*, 2nd edn. Edinburgh: Churchill Livingstone.

Gleeson, N. P. and Mercer, T. H. (1996) 'The utility of isokinetic dynamometry in the assessment of human muscle function', *Sports Medicine*, 21(1): 18–34.

Herzog, W. (1988) 'The relation between the resultant moments at a joint and the moments measured by an isokinetic dynamometer', *Journal of Biomechanics*, 21(1): 5–12.

Iossifidou, A. N. and Baltzopoulos, V. (1998) 'Inertial effects on the assessment of performance in isokinetic dynamometry', *International Journal of Sports Medicine*, 19(8): 567–573.

Iossifidou, A. N. and Baltzopoulos, V. (2000) 'Inertial effects on moment development during isokinetic concentric knee extension testing', *Journal of Orthopaedic and Sports Physical Therapy*, 30(6): 317–323; discussion: 324–317.

Kannus, P. (1994) 'Isokinetic evaluation of muscular performance: implications for muscle testing and rehabilitation', *International Journal of Sports Medicine*, 15(1), S11–18.

Kaufman, K. R., An, K. N. and Chao, E. Y. (1995) 'A comparison of intersegmental joint dynamics to isokinetic dynamometer measurements', *Journal of Biomechanics*, 28(10): 1243–1256.

Kaufman, K. R., An, K. N., Litchy, W. J., Morrey, B. F. and Chao, E. Y. (1991) 'Dynamic joint forces during knee isokinetic exercise', *American Journal of Sports Medicine*, 19(3): 305–316.

Kellis, E. and Baltzopoulos, V. (1995) 'Isokinetic eccentric exercise', *Sports Medicine*, 19(3): 202–222.

Kellis, E. and Baltzopoulos, V. (1996) 'Gravitational moment correction in isokinetic dynamometry using anthropometric data', *Medicine and Science in Sports and Exercise*, 28(7): 900–907.

King, M. A. and Yeadon, M. R. (2002) 'Determining subject-specific torque parameters for use in a torque-driven simulation model of dynamic jumping', *Journal of Applied Biomechanics*, 18(3): 207–217.

Maffiuletti, N. A. and Lepers, R. (2003) 'Quadriceps femoris torque and emg activity in seated versus supine position', *Medicine and Science in Sports and Exercise*, 35(9): 1511–1516.

Osternig, L. R. (1986) 'Isokinetic dynamometry: implications for muscle testing and rehabilitation', *Exercise and Sport Sciences Reviews*, 14: 45–80.

Pavol, M. J. and Grabiner, M. D. (2000) 'Knee strength variability between individuals across ranges of motion and hip angles', *Medicine and Science in Sports and Exercise*, 32(5): 985–992.

Perrin, D. H. (1993) *Isokinetic Exercise and Assessment*, Champaign, IL: Human Kinetics.

Reimann, U., Verdonck, A. J. and Wiek, M. (1997) 'The influence of a joint displacement on torque, angle and angular velocity during isokinetic knee extension/flexion, *Theoretical considerations*', *Isokinetics and Exercise Science*, 6(4): 215–221.

Tsaopoulos, D. E., Baltzopoulos, V., Richards, P. J., and Maganaris, C. N. (2011) 'Mechanical correction of dynamometer moment for the effects of segment motion during isometric knee-extension tests', *Journal of Applied Physiology*, 111(1): 68–74.

Tsirakos, D., Baltzopoulos, V. and Bartlett, R. (1997) 'Inverse optimization: functional and physiological considerations related to the force-sharing problem', *Critical Reviews in Biomedical Engineering*, 25(4–5): 371–407.

Yeadon, M. R., Jackson, M. I. and Hiley, M. J. (2014) 'The influence of touchdown conditions and contact phase technique on post-flight height in the straight handspring somersault vault', *Journal of Biomechanics*, 47: 3143–3148.

Yeadon, M. R., King, M. A. and Wilson, C. (2006) 'Modelling the maximum voluntary joint torque/angular velocity relationship in human movement', *Journal of Biomechanics*, 39(3): 476–482.

CHAPTER 9

DATA PROCESSING AND ERROR ESTIMATION

John H. Challis

INTRODUCTION

To paraphrase the English poet Alexander Pope (1688–1744),

To err is human; to quantify, divine.

Measured values differ from their actual (true) values; they contain errors. Quantification of the errors in any measurement should be made, so that the certainty with which a statement about the results of any analysis is known. Errors can arise during four stages of the experimental process, they are:

- Calibration
- Acquisition
- Data analysis
- Data combination

Calibration is the process of determining the parameters for a measurement device which gives the required output from the available input. Errors due to calibration arise from two sources: measurement system model and errors in calibration data. Calibration exploits knowledge of measurement system model input and desired model output to determine model parameters. No model is a perfect representation of the system so introduces errors. Errors due to calibration also arise because a perfect standard for calibration is not feasible; for example with a motion analysis system the control point positions, used for calibration, are not known to infinite accuracy. The process of calibration does not eliminate errors but should help keep them to a minimum.

During data acquisition, errors arise from a number of sources. These errors can arise because conditions change compared to when calibration was performed, for example the temperature of the sensors can change. During motion analysis, markers attached to the body segments are used to infer underlying bone motion, but the markers typically move relative to the bones

(Fuller et al., 1997). For time-series data sampled at an inappropriately low rate, the sampled data will contain errors.

Data analysis is the process via which sampled data are further processed to place the data in a usable form. One aspect to consider is that data are normally stored on a digital computer, which means that quantisation errors occur, as do errors due to finite computer arithmetic. It should be acknowledged that some data analysis procedures can reduce the errors corrupting the sampled data, for example the low-pass filtering of motion analysis data can have such an effect.

Data combination, the combination of parameters and variables, is necessary to determine many mechanical parameters or other descriptors of collected data. In these cases as these parameters and variables are combined, their errors propagate.

Good experimental practice is to seek to minimise the errors at their source, once this has been performed it does not matter what the source of the error is, but it is important to quantify the error. The chapter will outline the terminology used in error analysis, and provide examples of how errors can be minimised and determined.

DEFINITION OF KEY TERMS

A number of different terms are used when referring to error analysis; each term has its own specific meaning. To aid in reading the subsequent sections, in this section key terms are defined and discussed.

Accuracy

All measures contain errors; accuracy quantifies the bias in a measure caused by measurement error. Accuracy is quantified as the difference between a true value and an observed value. A measurement made with high accuracy will have little error, or a small bias from the true value. Accuracy is normally specified as the maximum error that can exist in a measurement but, in reality, it is inaccuracy that is being quantified. Accuracy is typically quantified by measuring the deviation between measures and a criterion.

Table 9.1 shows the output from a motion analysis system making ten repeat measures on a 1-meter long reference moved throughout the field of view of a motion analysis system. Accuracy can be assessed by measuring the bias between the reference length and its measured value. Three evaluation criteria are used: the mean, the absolute mean and the root mean square difference or error. They are calculated as follows:

$$\text{Mean: } \bar{x} = \frac{1}{n}\sum_{i=1}^{n}\Delta x_i \tag{9.1}$$

$$\text{Absolute Mean: } \bar{x}_{Abs} = \frac{1}{n}\sum_{i=1}^{n}\left|\Delta x_i\right| \tag{9.2}$$

$$\text{Root Mean Square Error: RMSE} = \sqrt{\frac{\sum_{i=1}^{n}(\Delta x_i)^2}{n}} \quad (9.3)$$

Where Δx_i is the difference between a measure and the criterion and n is the number of measures.

The mean value is –0.17 cm, but this is not a good measure of accuracy as negative and positive values can cancel one another out, compensating errors. The mean does provide some information, as the tendency is for the system to underestimate lengths. The absolute mean gives a much higher value for the system inaccuracy, 0.93 cm, as does the root mean square difference at 1.21 cm. Adopting the root mean square difference is recommended as it is the most conservative of the criteria. In this example, accuracy could be expressed as percentage of the field of view of the motion analysis system.

In this analysis, distances were determined by considering the distance between a pair of markers, so information from this pair of markers had to be combined to compute length. Therefore the measure of accuracy adopted here does not reflect just the system accuracy as it also includes any error propagated from the combination of data (see a description of this problem later in the chapter). Of course, a problem in the assessment of accuracy is finding a suitable criterion or absolute standard with which to compare measured values (Crease, 2011).

Table 9.1 Ten measures of a reference length measured by a motion analysis system throughout the calibrated volume.

Measure	*Criterion Distance (m)*	*Measured Distance (m)*	*Difference [Δ_{xi}] (cm)*
1	1.000	0.9849	–1.51
2	1.000	1.0106	1.06
3	1.000	1.0073	0.73
4	1.000	0.9720	–2.80
5	1.000	0.9964	–0.36
6	1.000	1.0142	1.42
7	1.000	0.9973	–0.27
8	1.000	1.0064	0.64
9	1.000	0.9982	–0.18
10	1.000	0.9962	–0.38
Mean Error			–0.17
Absolute Mean Error			0.93
Root Mean Square Error			1.21

Precision

If a system were measuring something that should remain invariant, deviations in the measured values would indicate a lack of precision. Precision is the difference between an observed and an expected mean value. It is typically quantified using the standard deviation of repeat measures of the same quantity:

$$\text{Standard Deviation } \sigma = \left(\frac{1}{n-1} \sum_{i=1}^{n} (x_i - \bar{x})^2 \right)^{1/2} \tag{9.4}$$

Precision and accuracy are sometimes confused, but are distinct quantities. Imagine trying to measure the point of peak pressure beneath the feet during standing using a pedobarograph. If the pedobarograph consists of one sensor only, the system will be very precise, but accuracy in the identification of pressure distributions will be very poor. Figure 9.1 illustrates the three possible permutations for combinations of accuracy and precision, when aiming at a target. Under certain circumstances, precision may be the more important characteristic of a measurement device, for example if looking at changes in joint angles because of some intervention.

Uncertainty

Any measurement will contain potential deviations from the true value, due to measurement system inaccuracy and system imprecision. The term 'uncertainty' is used to describe the potential deviation of a measure from its true value. Of course a happy coincidence might mean, for occasional measurements, that the measured value is the same as the true value; while this is seldom the case, it is feasible. Uncertainty can be stated by giving the range of values likely to include the true value: *measured value ± uncertainty*.

For many measurement systems, the imprecision is much greater than the system accuracy, so uncertainty is reported with the standard deviation as an estimate of the precision. In such cases the assumption is made that the uncertainty is normally distributed, in which case the true value lies within one standard deviation either side of the measured value 68.3 per cent of the time.

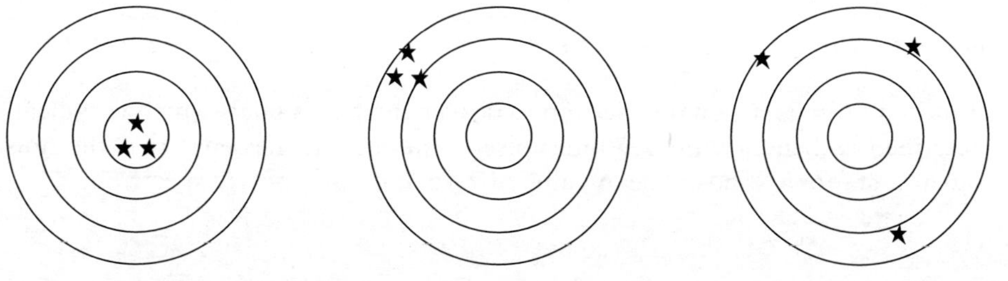

Figure 9.1 Three possible permutations for accuracy and precision, illustrated for shots at the centre of target. (a) High accuracy and high precision. (b) Low accuracy and high precision. (c) Low accuracy and low precision.

More conservative uncertainty estimates can be determined by using multiples of the standard deviation.

Resolution

For any measurement instrument there is a limit on the measured quantity which produces a change in the output of the instrument. Therefore, resolution reflects the 'fineness' with which a measurement can be made. Often resolution is expressed as a value or as a percentage of the instrument's full measuring range. If a measuring tape is used to determine segment perimeter length, then the resolution of the tape is indicated by the smallest sub-division on the tape. For example, if the tape is marked in 2 mm increments it may be reasonable to resolve measures to 1 mm. With measurement instruments interfaced to a computer, recorded values may have many digits but this does not necessarily reflect high resolution, as a large component of these values may reflect system noise.

Noise

Errors corrupting sampled data are referred to as noise. Noise is the unwanted part of sampled data; it masks the true value in which the experimenter is interested. Noise should be minimised in sampled data prior to further processing of the signal. The noise can either be systematic or random.

Systematic noise

Systematic noise is a signal superimposed over the true signal, which varies systematically and is correlated in some way with the measurement process, or the thing being measured. For example, the use of weights to calibrate a force plate could cause systematic error if their mass was 1 per cent less than that assumed during calibration. Many sources of systematic noise can often be modelled during the calibration process and removed from the sampled signal.

Random noise

Random noise is a signal superimposed over the true signal, which is typically described as being 'white'. White noise is random in the sense that the noise signal is stationary, has a mean value of zero and a flat power spectrum.

Propagated error

To perform mechanical analyses, different mechanical parameters have to be combined, so different sources of error impinge on the derived mechanical variable. Propagated errors occur when variables, which contain errors, are combined – the errors in the input variables cause errors in the derived variables. The section in this chapter on Combination of Variables and Parameters illustrates the mathematical principles behind error propagation. In the following sections, it is the intention to review the ways in which the uncertainties can be assessed and minimised for the major data collection and processing techniques used in biomechanics.

SAMPLING TIME-SERIES DATA

In many applications, data are collected at multiple time instants. The interval between samples may be months, as when measuring growth curves, or milliseconds if measuring centre of pressure motion during quiet stance. There are some basic principles that govern the collection of time-series data; these will be discussed in this section.

The sampling theorem gained widespread acceptance through the work of Claude Shannon; it states that a signal should be sampled at a rate that is at least twice the highest frequency component in the signal. This seems simple in concept but, before data collection, how does the experimenter know the highest frequency component of the signal? Previously published studies and pilot studies will provide useful guidelines, but if a participant exhibits unusual behaviour then this option may not be appropriate. A general guideline is to sample at a rate ten times greater than the anticipated highest frequency in the signal. Selecting such a high sample rate provides a safeguard if the analysed movement contains higher frequency components than normally expected. There are some additional advantages to a higher sample rate, which are outlined at the end of this section.

If the sampling theorem is not followed, the sampled signal will be aliased, which means that any frequency components higher than the Nyquist frequency (half of the sampling rate) will be folded back into the sampled frequency domain. Therefore, it is essential for the accurate analysis of time-series data that the signal be sampled at a sufficiently high sample rate, although it is not normal to know *a priori* the frequency content of a signal. For example, if signal 1 has a maximum frequency content of 1 Hz and is sampled at 8 Hz, and signal 2 has a maximum frequency content of 7 Hz, and is also sampled at 8 Hz, Figure 9.2 shows that signal 1 is adequately sampled, but signal 2 is not.

When sampling any signal, the experimenter is guided by Shannon's sampling theorem, which in essence states that any band-limited signal can be unambiguously reconstructed when that signal is sampled at a rate that is at least twice that of the signal band-limit. Slepian (1976) has shown that, theoretically, no signal is band-limited but, for practical implementation, the assumption is made that the signal is band-limited. Often overlooked in Shannon's work is the interpolation formula presented, which allows the sampled signal to represent the original analogue signal (Shannon, 1948). The use of a sample rate of only just greater than twice the band-limit of the signal allows accurate representation of the signal in the frequency domain, but not in the time domain. Figure 9.3 shows a signal with frequency components up to 3 Hz,

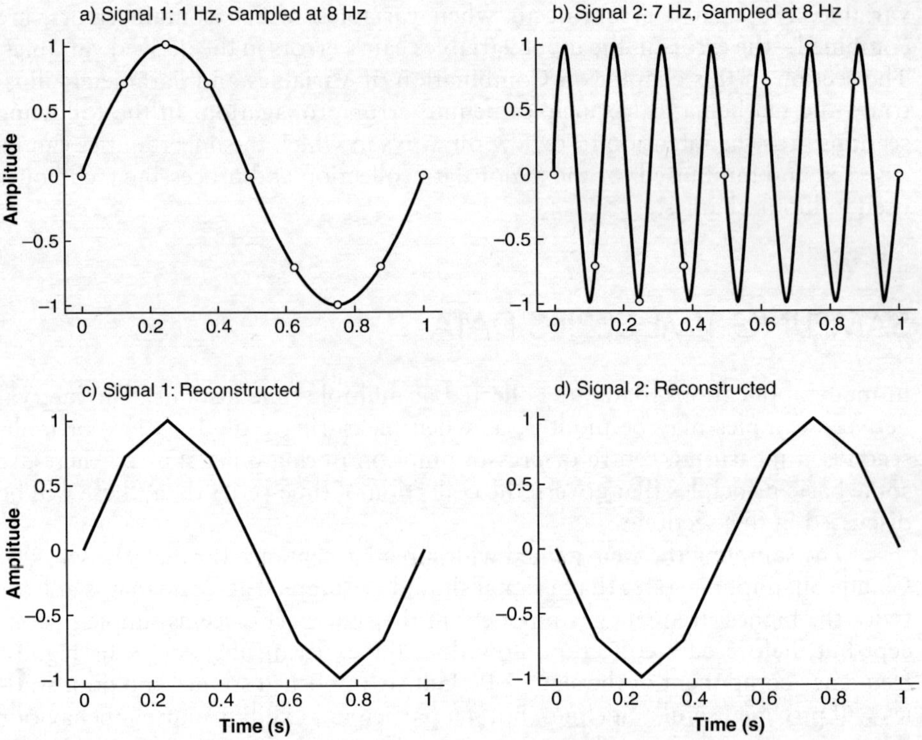

Figure 9.2 The influence of sample rate on reconstructed signal, where 'o' indicates a sampled data point.

which has been sampled at two different rates, 10 and 100 Hz. The 10 Hz sampled signal while representing the frequency components of the sampled signal, provides poor temporal resolution. For example, estimates of the minimum and maximum values of the signal in the 10 Hz sampled version of the signal produce an error of just under 6 per cent. The temporal resolution of the signal can be increased by interpolating the data; in Figure 9.3 the data sets have been interpolated using a cubic spline; now the signals at both sample rates are equivalent. Therefore, under certain conditions, further processing is warranted to produce from the sampled data a good representation of the signal in the time domain.

For certain applications, sample duration can be as important as sample rate. The lowest frequency that can be determined from collected data depends on the duration of the sample. The lowest frequency is called the fundamental frequency. For example, if a signal is sampled for 20 seconds, the lowest feasible frequency that can be represented in the signal is 0.05 Hz, that is, 1 / 20 s. If for a particular task the low frequency components of the signal are important, then sample duration should be selected appropriately.

In most analyses it assumed that random noise contaminating the data is white. One of the assumptions for white noise is that it is contains all

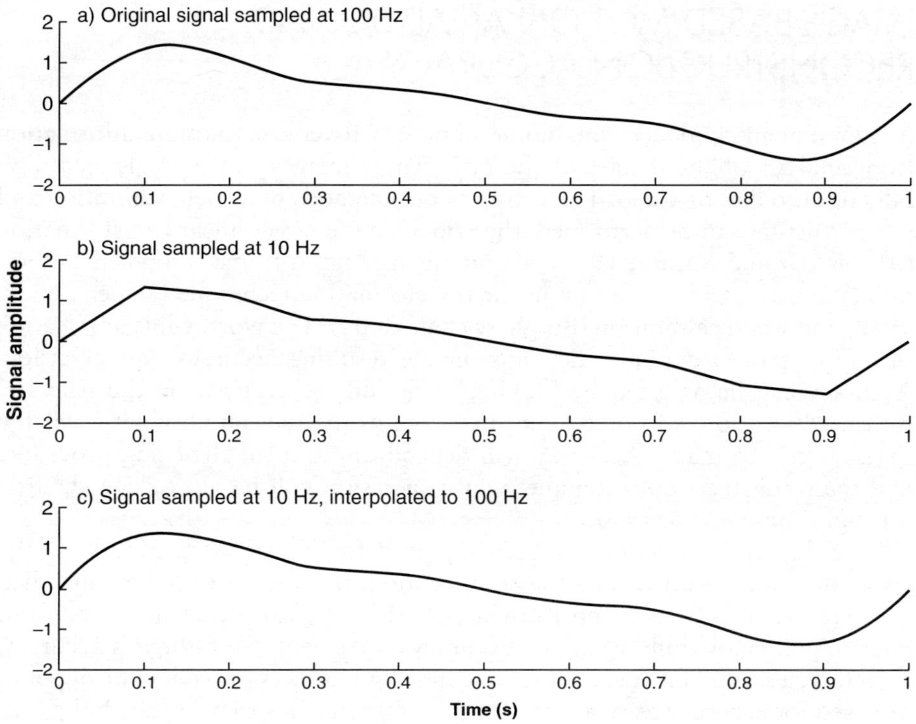

Figure 9.3 A signal with frequency components up to 3 Hz is sampled at two different rates, and then interpolated to a greater temporal density.

frequencies. Under this assumption some noise will always be aliased, as there is noise that exists beyond the Nyquist rate; this noise will be folded back onto the sampled domain. The appropriate sampling of a signal should also focus on adequate sampling of the noise. If noise is small compared with the signal, then a low sample rate may be appropriate as noise aliasing may not be a problem. If noise is large compared with the signal, then a high sample rate may be appropriate.

The following recommendations are made:

- Data should be sampled at a sufficiently high sample frequency to avoid signal aliasing.
- Signals should be sampled at a rate ten times greater than the anticipated highest frequency in the signal, as this provides a safety margin should the frequency content of the movement be atypical.
- After appropriate sampling of data, signal interpolation to provide greater temporal resolution should be considered.
- If low frequencies are of interest, sample duration should be selected appropriately.

IMAGE-BASED MOTION ANALYSIS AND RECONSTRUCTION ACCURACY

A common method used in biomechanics is to collect motion information from analysis of the images of the task. These images can be registered digitally or onto film or video. There are various methods by which calibration and reconstruction can be performed; these include the direct linear transformation (Abdel-Aziz and Karara, 1971), simultaneously multi-frame analytical calibration (Woltring, 1980), the non-linear transformation technique (Dapena et al., 1982) and wand calibration (Borghese et al., 2001). The worthwhile application of any of these techniques depends on the resulting accuracy and precision. Accuracy can be assessed by looking at the difference between the true and measured locations of points, and precision, as the repeatability with which a measure can be made. Accuracy will depend on both the digitising procedure and the reconstruction technique, whilst precision will be more dependent on the digitising.

Accuracy has commonly been reported for image-based motion analysis as the difference between the true location of the control points and their predicted values. Challis and Kerwin (1992) have shown that using the same control points for calibration and accuracy assessment overestimates accuracy; therefore, a second independent set of control points is essential if control points are used for accuracy assessment. There are alternatives to using control points to assess accuracy, for example a rod of known length can be moved throughout the calibrated space and the difference between its true length and predicted length compared. Wood and Marshall (1986) and Challis (1995c) have shown how the commonly used direct linear transformation technique produces less accurate predictions of the locations of points outside of the volume encompassed by the control points, compared with those inside this volume. These studies highlight the need to assess the accuracy of reconstruction throughout the space in which the activity takes place even if this is greater than the space encompassed by the control points.

It should not be assumed that because similar, or the same, equipment is being used as in other studies the reconstruction accuracy is the same. Many factors influence reconstruction accuracy and a small change in one may change the accuracy of reconstruction. For example, Abdel-Aziz (1974) showed how the angle between the intersections of the optical axes of the cameras causes different reconstruction accuracy.

Landmarks are either naturally occurring, such as a bony protuberance, or a marker of some kind is placed on the skin surface to specify the position and orientation of the segments with which they are associated (e.g. Donati et al., 2007). These landmarks or markers may not always be visible from all camera views, in which case their location has to be estimated. Skin movement can affect the ability of a marker to stay over the bony landmark it is intended to represent (Fuller et al., 1997). Assessment of inter-marker distance, for markers on the same segment, will reflect system accuracy, precision and marker movement.

When digitising manually, there are two types of precision: that of a single operator and that between operators. The main operator should digitise some sequences at least twice to get an estimate of precision, evaluated using the root mean square difference between the two measures. If more time is available, the operator should perform multiple digitisations of the same sequence, in which case precision can be estimated by computing the standard deviation. It could be that the operator is particularly imprecise, or is introducing a systematic error into the measurement process, for example by consistently digitising the wrong body landmark. Therefore, it is also recommended that a second operator digitise at least one sequence; this should quickly identify any systematic errors introduced by one of the digitisers, and it also gives a measure of inter-operator precision. Such an assessment analyses objectivity; if a low level of precision is evident between operators, then the assessment requires a high degree of subjective assessment to evaluate body landmark positions. Once precision has been quantified, the effect of this precision on variables computed from the position data should be evaluated.

Problems of marker location and identification can be particularly acute with automated image-based motion analysis systems for which, if a marker is obscured from a camera view, reconstruction may not be possible; and after this loss, the subsequent identification of markers may be problematic. Under these conditions either the operator is asked to make an estimate or the software attempts to make an 'educated guess' as to where the marker is. These influences should be assessed using the protocols just described. It should also be appreciated that with multiple camera systems a marker's position may be reconstructed from a variable number of cameras throughout an activity, resulting in changes in marker position accuracy.

The following recommendations are made:

- Reconstruction accuracy should be assessed using the root mean square difference between true control point locations and their estimated locations.
- The control points used for accuracy assessment should be independent of those used for calibration.
- If control points are not available, a rod of known length with markers on either end can be used for accuracy assessment, as long as it is moved throughout the calibrated volume.
- Analysing movements that occur outside of the calibrated volume should be avoided.

BODY SEGMENT INERTIAL PARAMETERS

For many biomechanical analyses it is often necessary to know the inertial parameters of the body segments involved, which, for biomechanical assessment, normally include the mass, moments of inertia, and the location of the centre of mass of each segment. Unfortunately, there are no ready measures of accuracy for individual segmental parameters. What can be assessed is the relative influence of different inertial parameter sets on other variables or parameters computed using the inertial parameters. For example Challis (1996) made three estimates of the moments of inertia of the segments of the lower limb; the resultant joint moments were then estimated at the ankle, knee and hip using each of the inertial parameter estimates. The percentage root mean square differences between the various estimates of the resultant joint moments were then assessed. For the two activities examined, maximum vertical jumping and walking, there was little difference between resultant joint moments, with the percentage root mean square differences all less than 2 per cent irrespective of which moment of inertia values were used. Such an approach has been extended by Challis and Kerwin (1996), to include all inertial parameters. This is not a direct assessment of accuracy but it facilitates evaluation of the influence of possible inaccuracies in inertial parameters. Given such an assessment it should be possible to conclude that the inertial parameters have been computed with sufficient accuracy for the analysis to be undertaken, if the influences of different inertial parameter estimates are within tolerable ranges.

Mungiole and Martin (1990) compared the inertial parameters estimated using different techniques with the results obtained from magnetic resonance imaging of the lower limbs of 12 males. These results are reassuring in that they show that different techniques all give very similar estimates of body segment inertial parameters. It could be decided to take one set and then perturbate that set by amounts which reflect the potential accuracy of that inertial parameter estimation procedure, rather than to take the measurements required to make different estimates of the segment inertial parameters. In this case, an estimate of the accuracy of the inertial parameters is required; here the data of Mungiole and Martin (1990) could be used for the lower limb. Yeadon and Morlock (1989) evaluated the error in estimating human body segment moments of inertia using both linear and non-linear equations, providing another source of data on which perturbations could be performed.

With certain methods for determining body segment inertial parameters, some measures of accuracy are possible. For example, if the method predicts the mass of the segments, these can be summed for all segments to give an estimate of whole body mass, which can be compared with actual body mass. Unfortunately, only one parameter is being assessed and compensating errors may mask genuine accuracy. When modelling the body as a series of geometric solids it is possible to compare model-predicted segment volume with actual volume (e.g. Hatze, 1980); volume is not normally required for biomechanical analyses, although it does a have a bearing on the accuracy of estimated masses.

The following recommendation is made:

- As there are no ready measures of accuracy for the inertial parameters of body segments, sensitivity analysis of these parameters should be performed.

LOW-PASS FILTERING AND COMPUTATION OF DERIVATIVES

For many biomechanical analyses it is necessary to reduce the white noise which contaminates the sampled data by low-pass filtering the data. For example, with image-based motion analysis data it is assumed the movement signal occupies the low frequencies, and the noise is present across all frequencies, to reduce the influence of noise the data are low-pass filtered. As white noise exists across the frequency spectrum some noise will still remain in the signal. There are a variety of methods used in biomechanics for low-pass filtering data, which can be placed into three broad categories:

- Digital filter, e.g. Butterworth filter (e.g. Winter et al., 1974)
- Splines, e.g. generalised cross-validated quintic spline (e.g. Woltring, 1986)
- Frequency domain based techniques, such as truncated Fourier series (e.g. Hatze, 1981)

To an extent, all of these approaches are equivalent (Craven and Wahba, 1979). After low-pass filtering the data appear smoother, thus low-pass filtering is sometimes referred to as smoothing.

In some applications it is also necessary to compute signal derivatives. In the frequency domain, signal differentiation is equivalent to signal amplification, with increasing amplification with increasing frequency. A simple example illustrates the influence of data differentiation on signal components. If the true signal is a sine wave with an amplitude of *1*, which is corrupted by noise across the spectrum, this example will focus on one noise component with an amplitude of 0.001. The signal to noise ratio can be computed for the zero, first, and second order derivatives:

Displacement data:	*x(t) = sin (t)*
Noise data:	*noise(t) = 0.001 sin (50 t)*
Signal to noise ratio:	*1: 0.001*
Velocity data:	*v(t) = cos (t)*
Noise data:	*noise(t) = 0.05 cos (50 t)*
Signal to noise ratio:	*1: 0.05*
Acceleration data:	*a(t) = -sin (t)*
Noise data:	*noise(t) = -2.5 sin (50 t)*
Signal to noise ratio:	*1: 2.5*

This example illustrates how important it is to low-pass filter data prior to data differentiation, to reduce the high frequency noise components of the sampled signal.

The question arises of how to select the degree of filtering or smoothing. If the properties of the signal and noise were known then the amount of the filtering could easily be selected, but this is not the case. If too low a cut-off frequency is selected the resulting signal will be incorrect as data from the original signal will have been discarded, if on the other hand the cut-off frequency is too high then too much noise will remain in the signal.

To specify the amount of filtering some researchers use previously published values, but this approach makes many assumptions about similarity of experimental set-up and subject performance across studies. It also assumes the original researchers made an appropriate selection. A range of procedures is available to automatically determine the amount of filtering. These approaches try to use information present in the sampled data to determine the amount of filtering; typically they assume that the noise is white. Hatze (1981), based on the work of Anderssen and Bloomfield (1974), presented an automatic procedure for using truncated Fourier series to low-pass filter and differentiate noisy data. Craven and Wahba (1979) presented a generalised cross-validation procedure for identifying optimal smoothing. This method fits a function to the data and determines the degree of fit, which minimises the root mean square fit to the data if each point in the data is systematically removed, and the fit is assessed by sequentially estimating these points. The generalised cross-validated quintic spline is most appropriate for the order of derivatives that are typically required in biomechanics (Woltring, 1985). To estimate the appropriate cut-off for the Butterworth filter, Challis (1999) exploited the properties of the autocorrelation of white noisc.

To illustrate some of the differences between these techniques, the performance of a Butterworth filter with the cut-off frequency selected using an autocorrelation based procedure (ABP – Challis, 1999), with a quintic spline with the degree of smoothing selected using a generalised cross-validation procedure (GCVQS – Woltring, 1986) were evaluated. Dowling (1985) presented data where he simultaneously collected accelerometer data and motion analysis displacement data. Figure 9.4 presents the estimates for the two techniques of the accelerometer data, along with the actual acceleration data. Notice that the ABP underestimates the peak acceleration values (Figure 9.4(b)), but performs better than the GCVQS around the region of limited acceleration (Figure 9.4(c)).

These automatic procedures are all, to some extent, black box techniques which should be used with caution. Their advantages are that they remove some subjectivity in assessing the degree of filtering, and as such can provide consistency between studies.

There is a case to be made for not using the same amount of filtering for all of the landmarks on a participant. Consider a gymnast rotating around a high bar; if the analysis focuses on wrist and ankle motion, the linear motion of the wrist will be much smaller than that of the ankle. The noise arising from a motion analysis system will be approximately equivalent for the wrist and the ankle, yet the amount of motion is different so they have different signal-to-noise ratios. Therefore, for this example it does not make sense to select the same amount of filtering for both landmarks. Similar situations exist for many activities: different landmarks have different signal-to-noise ratios, so require different amounts of filtering.

Lanshammar (1982) presented a series of formulae that provide an estimate of the noise which can be expected to remain in a noisy signal after smoothing and or differentiation. The basic format of the formula is:

$$\sigma_k^2 \geq \frac{\sigma^2.\tau.(2.\pi.\omega_b)^{(2k+1)}}{\pi(2k+1)} \tag{9.5}$$

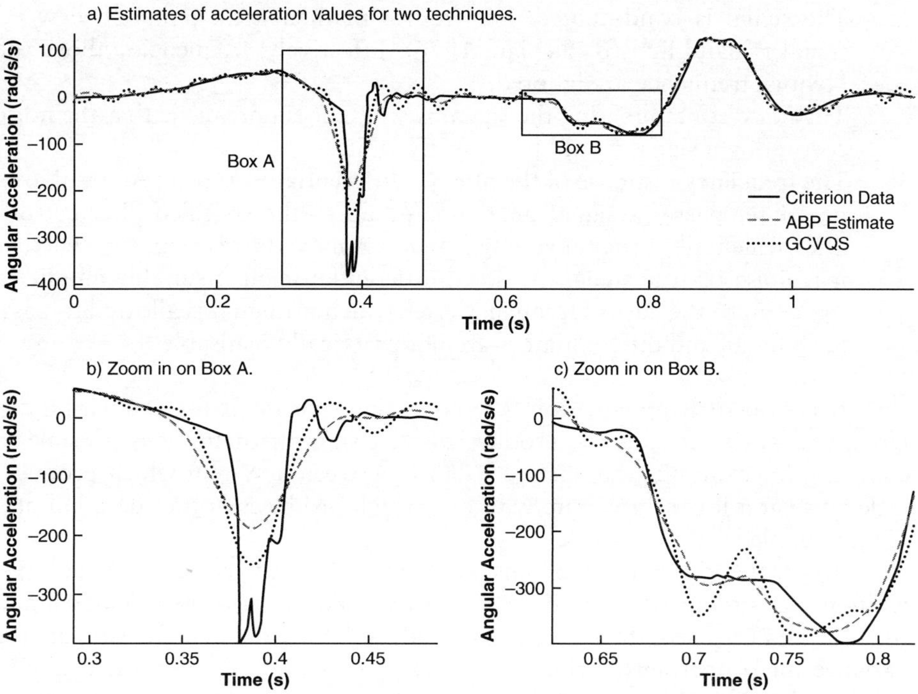

Figure 9.4 The performance of two filtering and differentiating techniques, ABP and GCVQS, for estimating acceleration data from noisy displacement data using criterion acceleration data of Dowling (1985).

Where,

σ_k^2 – minimum variance of the noise affecting the *k*th derivative, after processing,
σ^2 – variance of the additive white noise,
τ – sample interval,
ω_b – bandwidth of the signal in radians per second ($\varpi_b = 2.\pi.f_b$),
and f_b is the sample band-width in Hertz.

Equation 9.5 highlights a number of important points:

- Appropriate low-pass filtering of the data will reduce noise.
- Derivatives will always be of less precision than the original data.
- To increase data precision you have to decrease the variance of the additive white noise (σ^2), and or decrease the sample interval (τ), the latter effectively means increasing the sampling rate.
- Signal bandwidth is the final variable in equation 9.5, but is determined by the activity so cannot be adjusted by the experimenter.

There are three major assumptions associated with the series of formulae of Lanshammar (1982), which mean that the formulae give only minimum estimates. The assumptions are:

1 The signal is band-limited, but from a mathematical point of view no signal is band-limited (Slepian, 1976). Effectively, frequencies above the Nyquist frequency are ignored.
2 The noise contaminating the signal is white; it is unrealistic that the noise corrupting a signal would be perfectly white.
3 The frequency response of the filter or differentiator is ideal. An ideal low-pass filter passes a signal unattenuated up to the specified filter cut-off, after which all of the remaining signal is removed irrespective of whether it is noise or true signal. An ideal differentiator must suitably amplify a signal up to the cut-off frequency after which no signal is allowed to pass. Such filters and differentiators are not practically realisable.

It is evident from equation 9.5 that a figure of merit for comparing different analysis systems is the product $\sigma.\sqrt{\tau}$. Using this figure it is possible to compare, for example, the relative balance between a system which produces little noise at a low sample rate with one which produces noisier data but at a higher sample rate.

To use the formulae from Lanshammar (1982) the variance of the additive white noise is required. Even if derivatives are not going to be calculated and the raw data are used, the noise levels should still be estimated. To estimate the variance the properties of white noise can be exploited. If the noise contaminating a signal is white, this means the noise is not correlated between samples and has a mean value of zero, therefore on summing repeat measures the true underlying signal is additive, and the noise tends to its mean value. As a consequence taking the mean of a number of repeat samples, the signal to noise ratio is improved. If n repeat measures are taken, the true signals size is increased by a factor n, the variance of the noise is increased by the same factor, while the standard deviation of the noise is increased by a factor equal to the square root of n. On taking the mean, the signal amplitude to noise standard deviation ratio is increased by the square root of n. Such signal averaging has two applications: one as a method estimate signal noise, and the other to produce a noise-reduced estimate of a signal from repeat measures.

Winter et al. (1974) estimated noise levels by quantifying the variation in location of a marker on the foot when the foot was stationary during the stance phase of gait. Such an analysis does not provide information about the noise affecting all of the sampled data, and has limited application in certain activities. When quantifying the accuracy of derivative information if whole body motion has been analysed, the ground reaction forces can be estimated during ground contact (and then compared with force plate records), and the acceleration due to gravity during flight. Such analysis is not a direct way of assessing signal quality, as body segment inertial parameters are also required.

The following recommendations are made:

- Low-pass filtering of data is recommended, and is imperative if derivatives are to be computed.
- Different landmarks on the body will typically require different amounts of filtering.

- Selection of the amount of filtering should be prudently done, with automated procedures recommended.
- Estimation of the noise which can be expected to remain in a noisy signal after filtering and differentiation can be performed using the formulae of Lanshammar (1982).
- To compare motion analysis systems with different sample rates and noise levels, the following figure of merit should be used: $\sigma.\sqrt{\tau}$.

SEGMENT ORIENTATION AND JOINT ANGLES

When defining angles in either two- or three-dimensions, the rotations defined relate the transformation of one reference frame on to another. For example, segment orientation may be defined by relating an inertial reference frame to a reference frame defined for the segment. The transformation of coordinates measured in the segment reference frame to the inertial reference frame can be represented by

$$y_i = [\mathrm{R}]\,\chi_i + v \tag{9.6}$$

Where,

> y_i – position of point i on segment measured in inertial reference frame,
> [R] – attitude matrix,
> χ_i – position of point i on segment measured in segment reference frame,
> and v is the position of the origin of segment reference frame in the inertial reference frame.

Matrix [R] is a proper orthonormal matrix, which therefore has the following properties:

$$[\mathrm{R}]^{\mathrm{T}}[\mathrm{R}] = [\mathrm{R}]\,[\mathrm{R}]^{\mathrm{T}} = [\mathrm{R}]^{-1}\,[\mathrm{R}] = [I] \tag{9.7}$$

$$\det([R]) = +1 \tag{9.8}$$

Where,

> $[I]$ – identity matrix,
> and $\det([R])$ denotes the determinant of matrix $[R]$.

Once $[R]$ is determined, angles can be extracted from the matrix; for the two-dimensional analysis one angle, and for the three-dimensional analysis three angles. There are a number of procedures for determining this matrix, a comparison of a variety of approaches suggests the least-squares approach currently produces the most accurate results (Challis, 1995a). In a least-squares sense the task of determining $[R]$ and v is equivalent to minimising

$$\frac{1}{n}\sum_{i=1}^{n}\left([R]x_i + v - y_i\right)^T\left([R]x_i + v - y_i\right) \tag{9.9}$$

Where,

> n is the number of common landmarks measured in both reference frames ($n \geq 2$ for two-dimensional analyses, $n \geq 3$ and for three-dimensional analyses).

The computation techniques for minimising equation 9.9 include those presented by Veldpaus et al. (1988), Challis (1995b) and Challis (2001a) for two-dimensional analysis. Both the number of markers on a segment, and their distribution influence the accuracy with which equation 9.9 is solved, and therefore the resulting angles extracted from [R] (Challis, 1995b).

Given the locations of landmarks on the body of interest it is then common practice to determine the orientations of the body segments, and the relative orientation of adjacent segments (joint angles). In two-dimensions these segment orientations are normally computed by defining landmarks at the distal and proximal ends of the segment, the angle between a line joining these landmarks and the horizontal is then computed. It is possible to compute the precision with which these angles are measured using the following formula:

$$\sigma_\phi^2 = \frac{2.\sigma^2}{l^2} \tag{9.10}$$

Where,

> σ_ϕ^2 – error variance of angle,
> σ^2 – error variance of noise affecting coordinates,
> and l^2 is the distance between markers, or body landmarks.

The error variance of noise affecting coordinates can be estimated using the methods described in the previous section. The equation shows that to increase the accuracy of the angle computation, the noise affecting the coordinates should be minimised, and that the distance between the markers should be maximised. In the analysis of human movement there is a limit to the distance markers can be separated by on a segment, and marker locations must also be selected considering other factors such as minimising underlying skin movement, making sure the markers are visible to cameras, and possibly marking anatomically meaningful sites. These same principles apply to the determination of angles in three dimensions (Challis, 1995a). The error variance of the angle could also be arrived at by computing the angle a number of times from repeat measurements, and using this to establish the variance of the estimation.

The orientation of a segment or the orientation of one segment relative to another (joint orientation) in three dimensions is typically described by three angles extracted from the 3 by 3 attitude matrix. There have been recommendations for the definitions of segment angles and joint angles for the human body (Wu & Cavanagh, 1995; Wu et al., 2002; Wu et al., 2005), which in part relies on the appropriate specification of reference frames in the segments. The angles are normally defined as an ordered set of three rotations about a specified set of axes; in effect each angle is associated with a rotation about a given axis. For Cardan angles the resulting attitude matrix can be described in equation form, for example,

$$[R] = [R_x(\Theta_1)].[R_y(\Theta_2)].\ [R_z(\Theta_3)] \tag{9.11}$$

If a Cardanic angle sequence is selected and the three angles determined (Θ_1, Θ_2, Θ_3), then these angles might be associated with flexion/extension, abduction/adduction, and internal/external rotation. Variation in these angles can occur if a different sequence of rotations is selected, for example an x-y-z sequence compared with y-z-x sequence (Table 9.2). This means that to compare across participants and studies, the same reference frames must be defined in segments, and the same angle sequences used.

The error variances for three-dimensional angles are influenced in a similar way to angles in two dimensions. The errors in the three angles are highly correlated and cannot be considered independent when analysing signal noise; the covariance matrix for Cardan angles is

$$\text{var}\begin{pmatrix}\theta_1\\ \theta_2\\ \theta_3\end{pmatrix} = \sigma_\varphi^2 \begin{pmatrix} \dfrac{1}{\cos^2\theta_2} & 0 & \dfrac{-\sin\theta_2}{\cos^2\theta_2} \\ 0 & 1 & 0 \\ \dfrac{-\sin\theta_2}{\cos^2\theta_2} & 0 & \dfrac{1}{\cos^2\theta_2} \end{pmatrix} \tag{9.12}$$

Note that the influence of noise on the angles becomes greater when the middle angle approaches $\pm(2n+1)\ \varpi/2$, where n is any integer. At a middle angle of $\pm(2n+1)\ \varpi/2$ the other two (terminal) angles are undefined, the angle system has a singularity. The errors arising in these Cardanic angles increase as the middle angle approaches its singularity. The other commonly used system for defining angles is the Eulerian system (Craig, 2005), with the singularity for this angle system occurring when $\pm n.\varpi$ which is problematic as this occurs typically at the reference position (e.g. joint fully extended). Angle conventions and axes orientations should be selected not only to avoid singularities but also to keep the angles from approaching these singularities. Woltring (1994) has presented an angle set which avoids the singularities inherent in Euler or Cardan angles, where the eigenvector of the attitude matrix provides these angles – the so called helical angles. This convention does not have singularities and has a more favourable covariance matrix than the other angle conventions; the

Table 9.2 The influence of different angle sequences on the resulting amounts of rotations about each axis. This sequence effect is demonstrated for the six different Cardanic angle sequences.

Angle sequence	*Rotation about specified axis (degrees)*		
	X	*Y*	*Z*
X-Y-Z	60.00	5.00	−10.00
Y-Z-X	59.63	11.15	−0.72
Z-X-Y	57.70	21.21	−18.89
X-Z-Y	59.12	5.08	−9.96
Y-X-Z	59.62	9.93	−1.42
Z-Y-X	59.49	11.15	−0.73

disadvantage is that this angle definition lacks any physical significance. A similar critique can be levelled at other methods for parameterising the attitude matrix, such as Euler parameters and quaternions (Craig, 2005).

It should be noted that the error analysis presented in this section assumes that the error affecting the coordinates used to compute the angles is isotropic. This is not always the case, as some measurement systems have different resolutions in different directions, for example in three-dimensional analysis using video cameras the depth axis (axis perpendicular to a line between the principal points of a two camera system) is often measured with lower accuracy than the other two axes. Under conditions of anisotropic noise there may be increased error effects, although in most biomechanical applications noise is normally only mildly anisotropic. In addition the error analysis has assumed that the reference frames are accurately defined for the body segments, but this is challenging to accurately achieve (Brennan et al., 2011)

Sometimes helical or screw axes are used to quantify the motion at the joints. Although beyond the scope of this review, Woltring et al. (1985) presented formulae for the estimation of the error variance associated with estimating the helical axis parameters. If these variables are to be used to examine joint motion the researcher is directed to this work.

The following recommendations are made:

- As many markers as feasible should be used to define a segment.
- The markers should be distributed as far apart as possible, while avoiding areas where skin motion may be large.
- In three dimensions the angle definition system should be selected so that singularities, or approaching these singularities, are avoided for the movement to be analysed.

FORCE PLATES

Force plates are normally used to measure the ground reaction forces and the path of the centre of pressure. Hall et al. (1996) presented a procedure for the static calibration of force plates. The methodology allows calibration or confirmation of calibration for the three force and moment directions, with checks for cross-talk between axes. The methodology was used to evaluate a Kistler 9261A force plate, the results of this evaluation comfortingly confirmed the factory calibrations. Bobbert and Schamhardt (1990), by point loading, examined the accuracy of the centre of pressure estimates of a force plate. They were able to show that there were large errors in the centre of pressure estimates, particularly toward the edges of the force plate; they also presented equations to correct for these errors. Their analysis was for a larger model of force plate; the errors may be less in magnitude for smaller plates. Particularly in cases where the force plate is not installed in line with manufacturer's specifications, these types of checks should be mandatory.

Another less considered problem with the force plate is the natural frequency of the plate. Any object when struck will vibrate; the frequency of the vibration is often called the natural frequency. Many objects, particularly if they are composed of a number of different materials, have a number of frequencies at which they vibrate. It is normal to report a force plate's lowest natural frequency. The anticipation is that the frequency of the movement being analysed is lower than the natural frequency of the force plate, in which case the two can easily be separated. This appears to be the case for commercially available plates mounted according to manufacturer's specifications. If the amplitude of the natural frequencies is low then these can be ignored, if this is not the case they should be filtered and removed from the sampled data, ideally before the data are converted from an analogue to a digital form. Manufacturer's reported natural frequencies only apply if the force plate is mounted according to manufacturer's specifications (Kerwin and Chapman, 1988). If this is not the case the natural frequency of the plate should be examined, and then either the plate mounting modified, or the analogue signal appropriately high-pass filtered.

There is inherent variability associated with human movement (Newell et al., 2006), which raises questions about how many data trials should be collected to produce representative data. In this regard there have been a number of studies examining the repeatability of measures taken using force plates. These studies include those examining the influence of targeting and not targeting the force plate (walking – Grabiner et al., 1995; running – Challis, 2001b), and how many times the participant should run over the plate to get a representative ground reaction force pattern (e.g. Bates et al., 1983). While these factors are beyond the scope of this review, when designing a study using force plates, the experimenter is advised to consider these factors.

The analogue force data are converted into digital form for storage and analysis on a computer. The digitised values are each a single packet of information (word), represented in binary form [e.g. *10110* (binary) = *22* (decimal)]. Each word contains a certain amount of information (bits). This information varies with word length [e.g. *10110* (binary) contains *5* bits of information]; the more bits of information, the higher the resolution of the system. For example, if the input signal is *0* to *10* volts, and the analogue-to-digital converter has a resolution of *10* bits, then a single count would represent

$$\frac{10}{2^{10}} = \frac{10}{1024} = 0.0098\,volts \approx 0.01\,volts \qquad (9.13)$$

It is feasible to resolve to half a count; therefore, the *quantising error* is 0.005 volts. Depending on the signal to be analysed and the measuring equipment, quantising error may be very significant; see Figure 9.5 for an example.

To determine resultant joint moments, often information from a force plate and a motion analysis system are combined. Poor alignment between the force plate and motion analysis system reference frames can cause large errors in the computed moments (McCaw and Devita, 1995). Calibration procedures and calibration checking routines should be used to ensure these two reference frames are appropriately aligned (e.g. Rabuffetti et al., 2003).

The following recommendations are made:

- Confirm factory calibration of the force plate.
- If a force plate cannot be mounted to manufacturer's specifications, then the natural frequency of the plate should be determined.

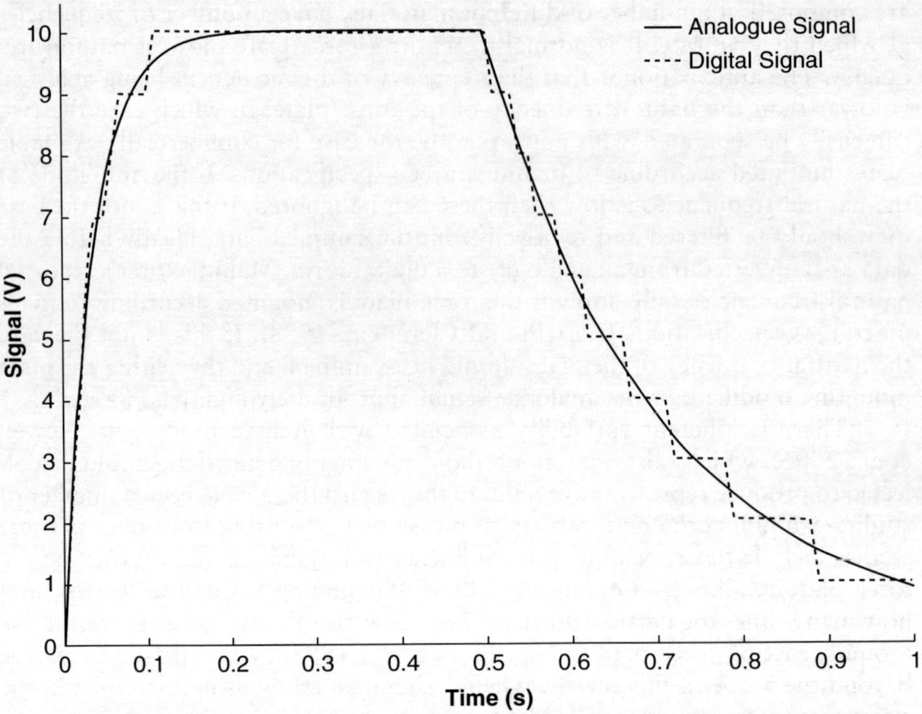

Figure 9.5 Example of quantisation error, where the resolution only permits resolution to 1 volt.

- An analogue filter should be considered for removing force components generated by the natural frequency of the force plate.
- The resolution of the analogue-to-digital converter should be matched to the required accuracy of the force measures.
- When the force plate is used in combination with a motion analysis system, it should be confirmed that the force plate and motion analysis system reference frames are aligned.

COMBINATION OF VARIABLES AND PARAMETERS

For any parameters or state variables there will be an uncertainty associated with their estimation; there will be propagation of these uncertainties in any value determined from their combination. The uncertainty in any value determined from the combination of parameters or state variables can be expressed mathematically (e.g. Barford, 1985). If the required state variable is P, which is a function of n state variables and or parameters then,

$$P = f\left(X_1, X_2, \ldots, X_n\right) \tag{9.14}$$

$$\left(\delta P\right)^2 = \sum \left[\frac{\partial P}{\partial X_i}.\delta X_i\right]^2 \tag{9.15}$$

Where,

X_i – state variable,
δP – uncertainty in the variable P,
$\frac{\partial P}{\partial X_i}$ – partial derivative of function P with respect to X_i,
and δX_i is the uncertainty in variable X_i.

Equation 9.15 shows how errors in the variables, or parameters required to compute another parameter or variable, propagate to produce an error in the computed value. For example if the function is the addition or subtraction of two variables, then

$$P = f(x,y) = x \pm y \qquad (9.16)$$

$$\frac{\partial P}{\partial x} = \frac{\partial P}{\partial y} = 1 \qquad (9.17)$$

$$\delta P^2 = \left(\frac{\partial P}{\partial x}.\delta x\right)^2 + \left(\frac{\partial P}{\partial y}.\delta y\right)^2 = \delta x^2 + \delta y^2 \qquad (9.18)$$

For the multiplication of two variables,

$$P = f(x,y) = x.y \qquad (9.19)$$

$$(\delta P)^2 = (y.\delta x)^2 + (x.\delta y)^2 \qquad (9.20)$$

Note that when combining two variables, it is not a simple case of adding or taking the mean of the errors in two variables to compute the error in the derived variable. Performing such an analysis assumes the uncertainties in the variables are random and uncorrelated. Equations 9.18 and 9.20 can easily be employed for simple biomechanical analyses, for more complex biomechanical analyses the resulting system of equations derived from equation 9.15 can become unwieldy.

Another way to assess the influence of these uncertainties is to perform a sensitivity analysis. The advantage of a sensitivity analysis are that it does not make the same assumptions as are required for equation 9.15, and that it is a potentially simpler approach when the analysis involves many variables. To perform a sensitivity analysis the parameters and state variables are changed (perturbated) by amounts relating to the estimated uncertainty or error with which they were measured. The changes in the required state variables as a consequence of this perturbation are then quantified. The major problem associated with assessing uncertainty in this way is how to allow for all the possible combinations of potential errors. Figure 9.6 shows the rectangular parallelepiped which contains all possible combinations of the errors, if for example the calculation of a key variable requires the combination of three variables (x, y and z). The figure illustrates that you theoretically have an infinite number of possible combinations of the errors. To some extent this problem can be viewed as a

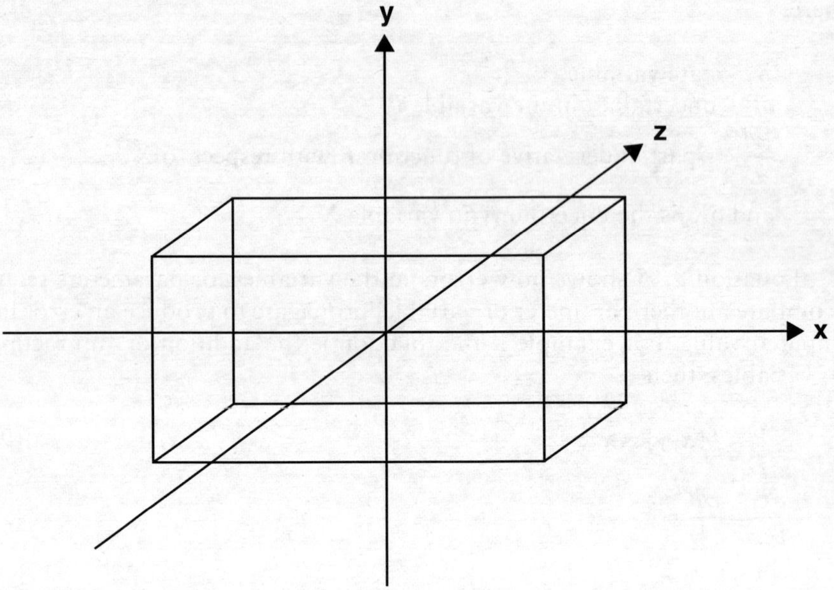

Figure 9.6 Graph showing the rectangular parallelepiped which encompasses all possible error combinations in variables x, y and z.

programming problem, as an appropriately written computer programme can easily investigate a reasonable sub-set of the options. Writing software for this purpose can be problematic and many commercial software packages are not written with this level of flexibility. A commonly used compromise is to run calculations for nine of the possible error options, one with the original data set with no noise (the origin in the figure), and each of the corners of the rectangular parallelepiped. The assumption here is that the larger the error the larger its resulting effect on any variable derived from using it, therefore considering the combinations using the maximum error levels examines the maximum error condition for the derived variable. In the majority of situations this is generally the case, although if time permits examining more than nine of the possible options is recommended.

If the uncertainty is quantified using either of the methods just discussed, then the problem of how to estimate the error in the variables and parameters still remains. The solution to this problem is to use a direct procedure where possible, for example procedures similar to those outlined earlier in this chapter, or to take an appropriate figure from the literature. Selection of values from the literature should be done prudently, by matching the selected published study to yours as closely as possible.

The following recommendations are made:

- Error propagation formulae can be used to quantify the error when two or more unrelated measures are combined.
- Sensitivity analyses can be used to estimate errors, and are particularly attractive if a large system of equations is being analysed.

CONCLUSIONS

There are many measurement procedures used in biomechanics, which are too numerous to cover in detail here. The preceding should have provided sufficient information to enable measurement accuracy to be assessed for other devices or procedures. Software for the processing of data, in the ways described in this chapter, is available from a number of sources. Basic signal processing software is available in MATLAB (www.mathworks.com/) and the freely available GNU Octave (www.octave.org/). More biomechanics specific code can be downloaded from the website of the International Society of Biomechanics (www.isbweb.org/).

The following broad recommendations are made:

- With any measurement procedure it is important to follow the basic experimental protocols, which means that the equipment must be accurately calibrated.
- Accuracy of calibration should be assessed using an independent measurement which assesses the accuracy throughout the range of required measurements.
- For time-series data, the sample rate should be selected to avoid signal aliasing.
- Error levels should be estimated and reported for all measurements.
- If the measurements are combined to derive further variables or parameters it is essential that the effect of combining errors from different sources is quantified and reported.

It is not possible to set acceptable levels of uncertainty, as these will vary from study to study. The purpose of the study dictates what the acceptable levels of uncertainty are, and a pedantic search for increased accuracy with a particular measuring device may be redundant for the purpose of one study, and yet be essential for another. Some of the factors which affect uncertainty levels cannot be known *a priori*, for example bandwidth of the sampled signal. When the measurement uncertainty cannot be sufficiently well estimated before data collection, a strict adherence to good experimental protocols followed by an error analysis is the experimenter's only recourse. Although the analysis of errors can be a tiresome process, it does give one the satisfaction of knowing what confidence can be placed on the results and any conclusions drawn from them.

REFERENCES

Abdel-Aziz, Y. I. (1974) 'Expected accuracy of convergent photos', *Photogrammetric Engineering*, 40: 1341–1346.

Abdel-Aziz, Y. I. and Karara, H. M. (1971) 'Direct linear transformation from comparator co-ordinates into object space co-ordinates in close range photogrammetry', in *American Society of Photogrammetry Symposium on Close Range Photogrammetry*, pp. 1–18. Falls Church, VA: American Society of Photogrammetry.

Anderssen, R. S. and Bloomfield, P. (1974) 'Numerical differentiation procedures for non-exact data', *Numerische Mathematik*, 22: 157–182.

Barford, N. C. (1985) *Experimental Measurements: Precision, Error and Truth*, New York: John Wiley.

Bates, B. T. L., Osternig, L. R., Sawhill, J. A. and James, S. L. (1983) 'An assessment of subject variability, subject-shoe interaction, and the evaluation of running shoes using ground reaction force data', *Journal of Biomechanics*, 16: 181–191.

Bobbert, M. F. and Schamhardt, H. C. (1990) 'Accuracy of determining the point of force application with piezoelectric force plates', *Journal of Biomechanics*, 23: 705–710.

Borghese, N. A., Cerveri, P. and Rigiroli, P. (2001) 'A fast method for calibrating video-based motion analysers using only a rigid bar', *Medical and Biological Engineering and Computing*, 39: 76–81.

Brennan, A., Deluzio, K. and Li, Q. (2011) 'Assessment of anatomical frame variation effect on joint angles: a linear perturbation approach', *Journal of Biomechanics*, 44: 2838–2842.

Challis, J. H. (1995a) 'An examination of procedures for determining body segment attitude and position from noisy biomechanical data', *Medical Engineering and Physics*, 17: 83–90.

Challis, J. H. (1995b) 'A procedure for determining rigid body transformation parameters', *Journal of Biomechanics*, 28: 733–737.

Challis, J. H. (1995c) 'A multiphase calibration procedure for the direct linear transformation', *Journal of Applied Biomechanics*, 11: 351–358.

Challis, J. H. (1996) 'The evaluation of the accuracy of human limb moment of inertias and their influence on resultant joint moments', *Journal of Applied Biomechanics*, 12: 515–529.

Challis, J. H. (1999) 'A procedure for the automatic determination of filter cutoff frequency of the processing of biomechanical data', *Journal of Applied Biomechanics*, 15: 303–317.

Challis, J. H. (2001a) 'Estimation of the finite center of rotation in planar movements', *Medical Engineering and Physics*, 23: 227–233.

Challis, J. H. (2001b) 'The variability in running gait caused by force plate targeting', *Journal of Applied Biomechanics*, 17: 77–83.

Challis, J. H. and Kerwin, D. G. (1992) 'Accuracy assessment and control point configuration when using the DLT for cine-photogrammetry', *Journal of Biomechanics*, 25: 1053–1058.

Challis, J. H. and Kerwin, D. G. (1996) 'Quantification of the uncertainties in resultant joint moments computed in a dynamic activity', *Journal of Sports Sciences*, 14: 219–231.

Craig, J. J. (2005) *Introduction to Robotics: Mechanics and Control*, 3rd edn. Reading, MA: Addison-Wesley Publishing Company.

Craven, P. and Wahba, G. (1979) 'Smoothing noisy data with spline functions: estimating the correct degree of smoothing by the method of generalised cross-validation', *Numerische Mathematik*, 31: 377–403.

Crease, R. P. (2011) *World in the Balance: The Historic Quest for an Absolute System of Measurement*, New York: W.W. Norton and Company.

Dapena, J., Harman, E. A. and Miller, J. A. (1982) 'Three-dimensional cinematography with control object of unknown shape', *Journal of Biomechanics*, 15: 11–19.

Donati, M., Camomilla, V., Vannozzi, G. and Cappozzo, A. (2007) 'Enhanced anatomical calibration in human movement analysis', *Gait and Posture*, 26: 179–185.

Dowling, J. J. (1985) 'A modelling strategy for the smoothing of biomechanical data', in B. Jonsson (ed.), *Biomechanics X-B*, pp. 1163–1167. Champaign, IL: Human Kinetics.

Fuller, J., Liu, L. J., Murphy, M. C. and Mann, R. W. (1997) 'A comparison of lower-extremity skeletal kinematics measured using skin- and pin-mounted markers', *Human Movement Science*, 16: 219–242.

Grabiner, M. D., Feuerbach, J. W., Lundin, T. M. and Davis, B. L. (1995) 'Visual guidance to force plates does not influence ground reaction force variability', *Journal of Biomechanics*, 28: 1115–1117.

Hall, M. G., Fleming, H. E., Dolan, M. J., Millbank, S. F. D. and Paul, J. P. (1996) 'Static in situ calibration of force plates', *Journal of Biomechanics*, 29: 659–665.

Hatze, H. (1980) 'A mathematical model for the computational determination of parameter values of anthropomorphic segments', *Journal of Biomechanics*, 13: 833–843.

Hatze, H. (1981) 'The use of optimally regularized Fourier series for estimating higher-order derivatives of noisy biomechanical data', *Journal of Biomechanics*, 14: 13–18.

Kerwin, D. G. and Chapman, A. E. (1988) 'The frequency content of hurdling and running', in *Biomechanics in Sport*, pp. 107–111, London: Institution of Mechanical Engineers.

Lanshammar, H. (1982) 'On precision limits for derivatives numerically calculated from noisy data', *Journal of Biomechanics*, 15: 459–470.

McCaw, S. T. and Devita, P. (1995) 'Errors in alignment of center of pressure and foot coordinates affect predicted lower extremity torques', *Journal of Biomechanics*, 28: 985–988.

Mungiole, M. and Martin, P. E. (1990) 'Estimating segment inertial properties: comparison of magnetic resonance imaging with existing techniques', *Journal of Biomechanics*, 23: 1039–1046.

Newell, K. M., Deutsch, K. M., Sosnoff, J. J. and Mayer-Kress, G. (2006) 'Variability in motor output as noise: A default or an erroneous proposition?', in K. Davids, S. Bennett and K. Newell (eds), *Movement System Variability*, pp. 3–22, Champaign, Illinois: Human Kinetics.

Rabuffetti, M., Ferrarin, M., Mazzoleni, P., Benvenuti, F. and Pedotti, A. (2003) 'Optimised procedure for the calibration of the force platform location', *Gait and Posture*, 17: 75–80.

Shannon, C. E. (1948) 'A mathematical theory of communication', *Bell Systems Technical Journal*, 27: 379–423.

Slepian, D. (1976) 'On bandwidth', *Proceedings of the IEEE*, 64: 292–300.

Veldpaus, F. E., Woltring, H. J. and Dortmans, L. J. M. G. (1988) 'A least-squares algorithm for the equiform transformation from spatial marker co-ordinates', *Journal of Biomechanics*, 21: 45–54.

Winter, D. A., Sidwall, H. G. and Hobson, D. A. (1974) 'Measurement and reduction of noise in kinematics of locomotion', *Journal of Biomechanics*, 7: 157–159.

Woltring, H. J. (1980) 'Planar control in multi-camera calibration for three-dimensional gait studies', *Journal of Biomechanics*, 13: 39–48.

Woltring, H. J. (1985) 'On optimal smoothing and derivative estimation from noisy displacement data in biomechanics', *Human Movement Science*, 4: 229–245.

Woltring, H. J. (1986) 'A Fortran package for generalized, cross-validatory spline smoothing and differentiation', *Advances in Engineering Software*, 8: 104–113.

Woltring, H. J. (1994) '3-D attitude representation of human joints: a standardization proposal', *Journal of Biomechanics*, 27: 1399–1414.

Woltring, H. J., Huiskes, R., de Lange, A. and Veldpaus, F. E. (1985) 'Finite centroid and helical axis estimation from noisy landmark measurements in the study of human joint kinematics', *Journal of Biomechanics*, 18: 379–389.

Wood, G. A. and Marshall, R. N. (1986) 'The accuracy of DLT extrapolation in three-dimensional film analysis', *Journal of Biomechanics*, 19: 781–785.

Wu, G. and Cavanagh, P. R. (1995) 'ISB recommendations for standardization in the reporting of kinematic data', *Journal of Biomechanics*, 28: 1257–1261.

Wu, G., van der Helm, F. C., Veeger, H. E., Makhsous, M., Van Roy, P., Anglin, C., Nagels, J., Karduna, A. R., McQuade, K., Wang, X., Werner, F. W. and Buchholz, B. (2005) 'ISB recommendation on definitions of joint coordinate systems of various joints for the reporting of human joint motion-part II: shoulder, elbow, wrist and hand', *Journal of Biomechanics*, 38: 981–992.

Wu, G., Siegler, S., Allard, P., Kirtley, C., Leardini, A., Rosenbaum, D., Whittle, M., D'Lima, D. D., Cristofolini, L., Witte, H., Schmid, O. and Stokes, I. (2002) 'ISB recommendation on definitions of joint coordinate system of various joints for the reporting of human joint motion-part I: ankle, hip, and spine', *Journal of Biomechanics*, 35: 543–548.

Yeadon, M. R. and Morlock, M. (1989) 'The appropriate use of regression equations for the estimation of segmental inertia parameters', *Journal of Biomechanics*, 27: 683–689.

CHAPTER 10

RESEARCH METHODS

Sample size and variability effects on statistical power

David R. Mullineaux and Jonathan Wheat

INTRODUCTION

The nature of most research in biomechanics involves the analysis of time-series data. In performing this analysis several sequential and linked stages of proposing, conducting, analysing and reporting of research need to be performed (Mullineaux et al., 2001). A key issue within each of these stages is proposed. First, the research question should be worthy, ethical and realistic, and used to attempt to outline how to propose, support or improve a theory. Second, the research design should be valid and use the best available methods. Third, the data analysis should be accurate and meaningful. Last, the research should be interpreted and reported effectively. In planning a study, attention is often devoted to assessing statistical power. Power is affected by the research design, which in turn influences the accuracy of the statistical data produced. The aim of this chapter is to highlight factors affecting power, both through the research design and implementation of data transformations, whilst also focusing on methods to explore and analyse time-series data.

STATISTICAL POWER

One of the major uses of statistics evident in the literature is in the form of 'statistical significance testing' of the null hypothesis. Essentially this application only provides the probability of the results when chance is assumed to have caused them (i.e. the null hypothesis is assumed to be true). Carver (1978, p. 383) refers to this as 'statistical rareness' where, for example, if $P<0.05$ then obtaining a difference this large would be 'rare' and expected to occur less than 5 per cent of the time as the null hypothesis is true. This single bit of information is used to identify whether the null hypothesis supports the data or not. Statistical significance testing does not provide any other information, such as a description, reliability, validity, generalisability, effect authenticity or importance of the data. Despite this limited information, there appears to be an over-reliance on 'statistical significance testing' in research that has prompted concern since early on in their use. Tyler (1931, p. 116) states 'differences which are statistically significant are not always socially important. The corollary is also true: differences which are not shown to be statistically significant may nevertheless be socially significant'. More recently, Lenth (2001, p. 1) states 'sample size may not be the main issue, that the real goal is to design a high-quality study'.

This recognised problem may suggest that 'statistical significance testing' should not be used. However, this is unwise, as scientific research is entrenched with the need for these statistics, and 'for scientists to abandon the textbook significance tests would be professional suicide' (Matthews, 1998, p. 24). Consequently, to reduce the impact of an over-reliance on statistical significance testing, the skill of the researcher is vital requiring that he or she considers statistical power through, for example: (1) carefully planning and implementing a research design; (2) using a variety of additional statistical tests to explore and describe data (e.g. descriptive, effect size, confidence intervals, standard errors, reliability); and (3) applying data transformations.

Statistical significance testing is encapsulated by the concept of 'statistical power', often simply referred to as 'power'. In essence, the research design and methodology affect power, and power affects the outcome of the analysis (i.e. statistical significance value). Power –'the ability of a test to correctly reject a false null hypothesis' (Vincent, 1999, p. 138) – is affected by factors in the research design, statistics and data, some of which have been summarised in Table 10.1.

Further to setting the factors in Table 10.1 prior to the study, the researcher is also able to apply a variety of solutions to the data obtained to help meet any assumptions. For instance, analytical solutions to replacing missing data, identifying and correcting for univariate and multivariate outliers, and correcting for violations of assumptions of linearity, normality and homoscedasticity can all be applied and would generally be expected to improve statistical power. A variety of these solutions are described in textbooks, where, for example, common transformations to improve the normality of different data distributions have been described by Tabachnick and Fidell (2001, p. 82).

In addition to corrections for assumptions, decreasing the variability in the data can increase statistical power. Although increasing sample size is cited the most for increasing statistical power, reducing variability in the data has been considered more cost-effective (Shultz and Sands, 1995). Estimating the sample size is simple, and it is often used to meet many ethics committees' requirement that power is appropriately set *a-priori* so that estimates of the Type I (i.e. probability of incorrectly rejecting a null hypothesis as it is true) and Type II (i.e. probability of incorrectly accepting a null hypothesis as it is false) error rates are acceptable for the planned study. Decreasing variability has solutions that are simple (e.g. using a more reliable measurement device or dependent variable)

Table 10.1 Research design, statistics and data factors affecting statistical power.

Factor	*Statistical power*	
	Lower	*Higher*
Research design		
Type I error rate, alpha (α)	Small (e.g. 0.05)	Large (e.g. 0.10)
Type II error rate, beta (β)[a]	Large (e.g. 0.40)	Small (e.g. 0.20)
Sample size, *n*	Small	Large
Reliability of measurement	Poor	Good
Statistic		
Design	Between	Within
Tailed	Two	One (but power zero if incorrect)
Distribution assumed test	Non-normal	Normal
Distribution free test	Normal	Non-normal
Data[b]		
Effect size, *ES*	Small	Large
Mean difference	Small	Large
Variability	Large	Small

[a]Typically set at four times α (Cohen, 1988) and used to calculate Power=$(1-\beta)*100$ (Eq. 1).
[b]Prior to the study estimates of *ES* determined from previous literature. In a simple form, *ES* is the ratio of the mean difference to variability.

and complex (e.g. applying a multiplicative scatter correction). An overview of methods to estimate sample size and decrease variability are provided in the following sections.

ESTIMATING SAMPLE SIZE

In recruiting participants for a study, the goal is to select a sample that is representative of the population. The sample statistics should therefore provide a reliable and unbiased predictor of the population parameter. In general, the probability of a biased sample increases as the sample size decreases; this increases the sampling error – the error in the estimate of a population parameter from a sample statistic. On this basis, tables have been developed to indicate the sample size (n) required for a given population size (N) that provide the variation, with known confidence, of the sample statistics from the population parameters (see Krejcie and Margan, 1970). Assuming that the results are to be based on a randomly selected sample and generalised to the population from which the sample was drawn, then, such tables are easy to use. On the basis of the three stages in sampling (i.e. define a population; identify a sampling frame;

use a sampling technique), identifying the sampling frame that provides a list of all members of the population and subsequently indicates the population size is sometimes straightforward (e.g. listing the population of the UK top 100 middle distance runners), but more often difficult (e.g. identifying the population size of runners experiencing lower back pain). In this former example, based on a table from Krejcie and Margan (1970) a sample of 80 of the UK's top 100 middle distance runners would be required to identify the population parameters to within 5 per cent with 90 per cent confidence. Consequently, the relatively small sample sizes used in biomechanics presumably have a much smaller confidence in their representation of the population parameters, which may offer a partial explanation for any variations in sample statistics found between similar studies. Nevertheless, often in a study the researcher is interested in assessing statistical significance, for example, whether differences between treatments can be attributable to chance at a specified statistical significance level. Consequently, alternative techniques are required to provide an estimate of the sample size when statistical significance is being used.

Selecting an appropriate sample size is principally related to achieving a desired statistical power. Methods of determining sample size with respect to statistical power are primarily based upon the interaction between four factors: n, ES, α and β (see Table 10.1). These methods include using statistical formulas, tables, prediction equations and power software. Overall, the uses of these methods are straightforward, examples of which are described in Table 10.2.

There are, however, a number of issues that can limit the usefulness of the methods for performing a power analysis. First, once a required sample size has been estimated, the practicality of implementing a study with a moderate or large sample size is difficult. In biomechanics research, often sample sizes are small owing to the complexity of biomechanical testing and often time consuming nature. In the 1998 *Journal of Applied Biomechanics* the sample size was reported to range from 3 to 67 (mean=14.5; Mullineaux et al., 2001). The ability to test greater numbers more efficiently is improving with technological advances, but large sample sizes will probably remain a rare phenomenon. This potentially introduces a bias of the sample not representing the population, and that statistical significance testing may have insufficient power thus increasing the Type II error rate. Nevertheless, often in sport and exercise biomechanics, as the research question applies to smaller populations (e.g. technique of elite performers) this may introduce a smaller bias than in research applicable to larger populations (e.g. ergonomics of industrial workers), and the high reliability of many of the measurement techniques will increase statistical power.

Second, these methods generally rely on *ES* being determined from previous literature, and using criterion values denoting what is a small, moderate or large effect. For example, effect sizes of 0.2, 0.5 and >0.8 represent small, moderate and large differences (Cohen, 1988), respectively, where a moderate difference is considered visible to an experienced researcher (Cohen, 1992, p. 156). Both determining *ES* from previous literature and using criterion values for effect size have been considered a major limitation of estimating power (Lenth, 2001). This is emphasised by Mullineaux et al. (2001, p. 749) in that the effect size should consider 'what is an important difference theoretically, ethically, economically and practically, and considering whether the differences are

Table 10.2 Examples of the use of four methods of power analysis for estimating sample size.

Statistical formulas exist for some common tests, but not all. For the Independent *t*-test, $n=(2s^2(z_\alpha+z_\beta)^2)/\Delta^2$ (Eq. 2, Vincent, 1999, p. 142). The variables are: *s* is the estimated population standard deviation; Δ is the estimated difference between the two group means; and Z_α and Z_β are the *Z* scores on the normal curve for the desired alpha and beta values, respectively. To determine Z_α and Z_β, tables of the area under the normal curve can be used, or more easily use the NORMSINV function in Excel (Microsoft Corporation, Redmond, WA). For example, the sample size for a two-tailed independent *t*-test, where *s*=8, Δ=7, α=0.05 and β=0.20, using *n*=(2**s*^2*(NORMSINV(α/2)+NORMSINV(β))^2)/Δ^2 (Eq. 3) indicates 20 participants per group are required to achieve a power of 80% (i.e. (1–β)*100; Eq. 1). Note, in equation 2, α is divided by 2 for the two-tailed test – if a one-tailed test was preferred then the required sample size would be 18 participants per group.
Tables, based on *n*, *ES*, α and β, have been developed for many tests. For example, based on effect sizes of small (*f*=0.10), moderate (*f*=0.25) and large (*f*=0.40), and reading the table from Cohen (1988, pp. 315–6), to achieve an 80% power for a 1-way between ANOVA for 4 groups, α=0.05 and a moderate effect size (i.e. *f*=0.25), the required sample size is 44 per group.
Prediction equations exist for several tests. These formulas are particularly helpful for repeated measures designs where formulas and tables are generally unavailable. For instance, Equation 4 can be used to determine the power (Ω) for the group main effect for a two group mixed ANOVA with α set to 0.05 (Park and Schutz, 1999, p. 259, where: *n* is sample size; *d* is effect size; *r* is mean correlation between the repeated measures; *q* is number of levels of the repeated measures factor). Setting up a spreadsheet to use these formulas is beneficial. Hence, to approximate the power to 0.80 (i.e. 80%) for *d*=0.6, *r*=0.5 and *q*=3, the required sample size is 30. Note, the powers are different for the repeated measures factor (85%) and the interaction (50%), and these are determined using additional formulas. $\Omega = -0.00000442n^2 + 0.115/d^2 + 0.085\sqrt{n} - 0.785\sqrt{r} - 0.917/d - 2.066/\sqrt{n} - 0.329/q + 2.585$ (Eq. 4)
Power software is available, including some interactive websites. For example, using the web page www.math.yorku.ca/SCS/Online/power/ (Friendly, 2006) and the example for the Table method (i.e. where 44 participants were required per group for a 1-way ANOVA, 4 groups, α=0.05 and moderate effect size), from this website an estimate of 40 participants per group are required. The website is set up as follows: • Number of levels of effect 'A'=4 (i.e. the 4 factors of the one-way between ANOVA). • Total number of levels of all other factors crossed with effect 'A'=1 (i.e. 1 for a one-way ANOVA). • Error level (alpha) for which you want power or sample size calculated=0.05. • Effect size(s) for which you want power or sample size calculated=0.25, 0.75, 1.25 (note, in this website these three values correspond to effect sizes of small, moderate and large).

bigger than limits identified for error, reliability and variability analyses'. This is further supported by Hopkins et al. (1999) whom suggest that differences between elite performers are tiny and, as such, much smaller effect size values might be meaningful in such situations. Alternatively, the researcher might decide on a particular effect size for the study being undertaken. For example, if the findings are to have real practical implications, then a large effect size of 0.8 might be chosen. But if the research is new, and the researcher feels it is important to detect any difference, even if small, then an effect size of 0.2 could be chosen (cited *ES* values based on the definitions of Cohen, 1988).

Third, *ES* is affected by many factors, some general ones of which are described in Table 10.1. For example, on-line motion capture systems are more reliable, although not necessarily more valid, than manual digitising. Greater reliability reduces variability, which in turn will increase the *ES* measure. Consequently, estimates of *ES* from previous literature should factor in differences in methodologies, or acknowledge or recognise limitations in the work. A further problem could simply be that the information to estimate *ES* does not exist, as no similar variable has previously been used or reported. Reporting means and standard deviations in abstracts for key variables is thus important as these provide the reader with data for a simple estimate of *ES*.

Although power is considered important by many authors (e.g. Howell, 1997), setting the sample size based on other issues both increases power and meets other statistical assumptions. For example, using a sample size of no less than 30 has been recommended per group when testing for differences between groups (Baumgartner and Strong, 1998). This is to increase the probability of a normal sampling distribution on the dependent variable, an assumption on which statistical tables, such as those of the t and F-statistic, are based. In addition, specific statistical tests benefit from minimum sample sizes to improve the validity of the analysis. For example, a small sample size (n) to the number of predictor variables (p) ratio in regression analysis can cause problems. A small ratio can result in high correlations occurring by chance and the generality of the regression equation will be limited. The recommended n:p ratio varies between authors, hence Mullineaux and Bartlett (1997) propose using at least 10:1 and ideally have between 20:1 and 40:1, where the largest ratio would be required for a stepwise multiple regression.

In addition to determining the sample size required on the basis of statistical grounds, consideration of a number of other factors is important. For instance, estimating and minimising the attrition rate of participants from, for example, injury, absence or dropout should be made. As well as losing participants during the study being time consuming, it is also problematic, as these participants may possess the principal characteristics that you may be most interested in studying. For example, in analysing the propensity of elderly participants to hip fractures, the increased risks observed in some testing procedure may cause hip fractures and subsequently a loss of the participant. Consequently, it is important to implement a design with some compromise over the 'desired' protocol to minimise attrition. Reducing the intensity of the protocol, number of testing sessions or duration of the experiment may be suitable. Also, it is important to provide an adequate extrinsic motivation (i.e. explain 'what is in it' for them) and to interact with the participants. Hence possible methods to reduce attrition include sampling participants carefully, keeping participants informed, collecting data efficiently, acting courteously and professionally at all times, talking with clients, making the testing interesting, explaining the results, minimising potentially embarrassing environments (e.g. do not test in a publicly visible laboratory) and keeping a clean and tidy laboratory. However, some participant attrition, missing data or removing outlier scores is probable. The resulting missing data should be evaluated, first for patterns (e.g. outlier scores may be associated with a particular trait and are therefore important) and then to whether the data should be replaced. Replacing these values is advantageous as it retains participants in the data set, and consequently the degrees of freedom remains larger thus increasing statistical power. In contrast, there are always limitations in replacing missing values. These will not be covered here, although using prior knowledge, mean scores, regression analysis, expectation maximisation and multiple imputation are explained as methods of replacing missing data (see Tabachnick and Fidell, 2001, p. 60). Another simple method is that the missing value should equal the row mean plus column mean minus grand mean (Winer et al., 1991). In terms of outliers, there are a variety of one-dimensional (i.e. spatial) approaches where a commonly cited definition is a value exceeding ±3 SD away from the mean. However, as parametric-based statistics can be ineffective in detecting outliers, it has been proposed that outliers are better identified as values exceeding ± 2.5 × 1.48 MAD (Leys et al., 2013) where MAD is median absolute differences and 1.48 approximates MAD to SD). For two-dimensional data (i.e. spatial and temporal) that are appropriate for time-series data, there are a variety of outlier detection methods that can be applied, including 'window' (Mullineaux and Irwin, 2017) and 'shape' (Arribas-Gil and Romo, 2014; Sangeux and Polak, 2014) based approaches. It should be cautioned, however, that regardless of identifying any outliers, these should not be removed without a clear rationale. Where there are obvious errors, removal is clearly justified, although in other instances, such as where the functionality of variability is being explored, their removal may be less clear and may contravene the theoretical assumptions of the analysis.

In summary, if sample sizes are small then generalising the results to a larger population should be viewed with caution. Where statistical significance testing is being used to determine whether chance can account for the difference, then power analyses are beneficial in choosing an appropriate sample size, but power analyses must not be used *post-hoc* to help explain results (Lenth, 2001). Power analyses can be problematic, however, in that they require data from previous literature that may not necessarily be of adequate quality or even address variables of interest. In addition, these analyses principally require sample size to be altered, whereas other factors influencing statistical power can be used, some of which are described in Table 10.1. These may provide the means to implement more efficient experimental designs (e.g. powerful, yet without large sample sizes). Such additional factors to increase statistical power are considered in the next section.

REDUCING VARIABILITY

In biomechanics, large variability often exists between individuals, which can arise through the methods and the population being investigated. Reducing this variability provides a key method in increasing statistical power. Variability can be reduced through the research design and methods by, for example, using a more reliable measure, averaging several trials, removing noise through filtering or correcting for the violation of statistical assumptions. In addition, reducing variability by catering for the individuality in the data is also effective which may be particularly important in 'elite' and 'injured' populations owing to their uniqueness and small sample size that make generalisations difficult. A principal method of removing systematic differences between participants is by normalising the data either as a ratio or through an offset. Alternatively, if inferential statistics are not required then simply providing descriptive statistics is an effective method of reporting the results as has been provided in past (e.g. Burden et al., 1998). Several of these approaches will be expanded upon.

Ratio normalisation

Ratio normalisation is underpinned by the assumption that a theoretical and statistical relationship exists between the dependent variable and a covariate. When this relates to an energetic variable as the dependent variable and a body dimension as the covariate then normalisation is a form of scaling known as allometry (Schmidt-Nielsen, 1984). Applying allometry is generally beneficial as it is able to cater for non-linear relationships, although the result may be linear. There has been a renewed interest in this area applied to humans, but scientific literature on allometry dating back to 1838 has been cited (see Winter and Nevill, 2001). Generally, allometry is appropriate for identifying the extent to which performance differences are attributable to differences in size or to differences in qualitative characteristics of the body's tissues and structures (Winter and Nevill, 2001, p. 275). In addition, as allometry can cater for non-linear relationships, it is suitable for addressing the non-isometric and isometric changes in dimensions with growth (Tanner, 1989). It is important that such analyses are underpinned by a theory. Dimensionality theory offers one possible theoretical basis for an allometric scaling analysis.

Dimensionality theory is underpinned by the Système International d'Unités that comprises seven base units (mass, length, time, electric current, temperature, amount of substance and luminous intensity). From these seven units all other units can be derived (e.g. area, volume, density, force, pressure, energy, power, frequency). Often for convenience, these units are renamed, such as, the units of force of $kg.m.s^{-2}$ are denoted as newtons (N). Dimensionality theory can be used for two main purposes (Duncan, 1987): dimensional homogeneity and dimensional analysis. Dimensional homogeneity can be used to check that an equation is correct by partitioning both sides of an equation into their base units. If the units on both sides are partitioned into the same base units then the equation is correct. Dimensional analysis can be used to predict the relationship between different dimensions as a means to provide a theoretical foundation for a research study, such as, between several continuous variables.

When there are two continuous variables (y and x), they can be scaled with each other in many forms, three of which are common: ratio standard (y=bx), linear regression (y=a+bx) and non-linear (y=ax^{b}), where a and b are constants. The use of each of these scaling techniques should be dictated by theory, although the theory is not always obvious. In addition, the mathematics of, and the assumptions for, a scaling technique delimit their use theoretically and statistically. For example, non-linear scaling may be appropriate for data that are not necessarily linear, theoretically requires a zero intercept and contain multiplicative error about the regression. The opposite assumptions are made in ratio standard or linear regression scaling analyses in that the data are linear, the intercept is not fixed at zero and the error is additive. Many of these assumptions are explained in textbooks (e.g. Tabachnick and Fidell, 2001) and are illustrated in a paper where these assumptions also underpin some repeated measures reliability statistics (Mullineaux et al., 1999).

Another benefit of scaling is that it provides the data normalisation that removes the effects of a covariate. An additional benefit is that more than one independent variable, including one dummy variable (i.e. a dichotomous independent variable coded as 0 and 1), can be included in a multiple scaling analysis. This will increase the explained variance and reduce the effect of extraneous variables, but a larger number of statistical assumptions (e.g. multicollinearity) need to be checked to validate the analysis. This has been applied in an allometric scaling analysis to determine a gender difference in left ventricular heart mass (e.g. Batterham et al., 1997). Although simple scaling is easy to perform (e.g. $y=ax^{b}$), when there is more than one independent variable (e.g. $y=ax_1^{b}+cx_2^{d}$, where x_1 and x_2 are independent variables, and c and d are constants) in non-linear scaling it is easier to use a log-log transformation method in combination with multiple linear regression analyses. As the data are non-linear, where the error is often multiplicative, then the first log transformation generally linearises the relationship, alters the error to being additive and improves the normality distribution of the data. These are all necessary assumptions underpinning a linear regression scaling analysis.

In scaling analyses, providing a theory that supports the relationship identified can be difficult. In particular, this is difficult in data analyses that routinely use ratio standard or linear regression scaling analyses and yet their use is still common. For example, normalisation to: lean body mass for ground reaction forces (e.g. Raftopoulos et al., 2000), limb length for stride length (Cham and Redfern, 2002), and maximal isometric contractions for electromyographic data (e.g. Pullinen et al., 2002) have been used. A further limitation of ratio standard and linear regression scaling has been proposed in that extrapolation beyond the actual data range should be avoided as this would not generally meet the zero intercept assumption (Batterham et al., 1997). However, this caution may also need to be applied to allometrically scaled relationships as the assumption of a zero intercept is often beyond the data range that could be tested empirically. A further consideration is that although scaling is often used to remove the effect of a covariate (x) on the dependent variable (y), alternative methods are available. Nevertheless, non-linear and specifically allometric scaling has been demonstrated to be superior to both ratio-normalisation and ANCOVA in removing the effects of a covariate on the dependent variable (Winter and Nevill, 2001). To confirm that the effects of a covariate have been removed through normalisation, a correlation between the dependent variable and the newly normalised data should be close to zero (Batterham et al., 1997). If this correlation is not near zero, or at least is not smaller than the correlation between the dependent variable and the covariate, then the normalisation has been inappropriate and may have introduced more variability into the data that will decrease statistical power.

One method of providing theoretical support for a suitable scaling factor is to use dimensionality analysis. For instance, force is proportional to mass, mass is proportional to volume, and volume to the power of 0.67 equals cross-sectional area (CSA). Consequently, CSA and force are proportional to mass to the power of 0.67 (see surface law explanation in Schmidt-Nielsen, 1984). Empirically, using dynamometry, it has been recommended that muscle force be normalised to body mass to the power of 0.67 (Jaric, 2002). In another example, to cater for participants of different sizes often GRF (ground reaction force) are normalised to body mass (e.g. Duffey et al., 2000), although it has been suggested that normalising to lean body mass is more appropriate (Raftopoulos et al., 2000). A comparison of normalising GRF and loading rates to body mass to different power exponents has been performed (Mullineaux et al., 2006). It was reported that GRF were appropriately normalised using body mass to the power of 1 (i.e. linearly), but that for loading rates a power exponent of 0.67 was generally better. Consequently linear normalisation of loading rates may inappropriately favour heavier participants because the proposed 0.67 power exponent would result in a greater relative reduction in loading rates for heavier than for lighter participants. However, the correct allometrically scaled relationships between energetic variables (e.g. GRF, muscle force) with body dimensions (e.g. body mass, CSA) requires further investigation.

The ratio normalisation methods mentioned earlier can be classified as a 'simple ratio' type. More complex methods also exist which include standard normal variate transformation (i.e. normalise to the standard deviation of the data) and derivative normalising (i.e. normalise to the magnitude of the first derivative of the data). The 'simple ratio' methods change the magnitude of the data, but the complex methods may also change the shape of the time-series data. Details on these complex methods will not be covered further. In the next section, an alternative form of normalisation using offsets is covered. Some of the benefits and limitations that are further expanded upon with respect to offset-normalisation below may well be appropriate to consider with the ratio-normalisation techniques covered here.

Offset normalisation

Normalising data using offsets can also be classified as simple and complex. Simple offsets commonly include subtracting the difference between the individual and the group mean for a defined measure. The measure may include a value within the data (e.g. initial, key instant, minimum, mean) or external (e.g. morphological value, criterion value). More complex methods include detrending, Fourier transform and multiplicative scatter correction. Detrending involves fitting a polynomial regression to the data, which is then used as the defined measure. The Fourier transform, that is, using the waveform of sum of weighted sines and cosines as the defined measure, has been used to correct skin marker movement in kinematic analyses of horses (van Weeren et al., 1992). The use of multiplicative scatter correction, using the linear regression coefficients of the mean data as the defined measure, has been demonstrated for reducing variability in kinematic data (Mullineaux et al., 2004).

As with all techniques, offset normalisations can be used in a variety of ways. Multiplicative scatter correction contains additive (intercept of the regression) and multiplicative (slope of the regression) elements that can individually or, as recommended, both be removed (Næs et al., 2002). Generally offset normalising reduces the variability in the data, and the shape of the data is retained with the simple offsets, but the shape is often altered with more complex methods.

In biomechanics, the use of offset normalising may be beneficial, particularly for kinematic data, due to variations in morphology that produce differences in limb lengths and standing angles. Correcting for some of these variations in morphology and kinematics may be important partly to improve statistical power, but also to assist in a valid interpretation of the data. In equine biomechanics, some simple offset normalisation techniques have been applied to cater for variations in morphology, including using a joint's standing angle (e.g. Back et al., 1994), angle at hoof impact (e.g. Holmström et al., 1994; although this was incorrectly referred to as a multiplicative scatter correction) and mean trace (e.g. Degueurce et al., 1997).

Assuming the data have been normalised to the same length in time of 101 points, the simple offset formulae is $New_{ki}=Original_{ki}+(D_{MEAN}-D_k)$. Hence, to obtain the simple offset normalised data (New_{ki}), the non-normalised data ($Original_{ki}$) for each participant (k) at each percentage of time (i) needs to be summed with the difference between the defined measures of the group mean (D_{MEAN}) and each participant's mean (D_k). As an example of a complex offset normalisation, the formula for a multiplicative scatter correction (MSC) is $MSC=Original_{ki}-a_k/b_k$, where the linear regression for the entire stride ($y_k=a_k x+b_k$) between the mean group (x) and each participant (y_k) provides the constants a_k and b_k (Næs et al., 2002).

Comparison of three simple offsets (mean, initial and standing angles) and one complex offset (multiplicative scatter correction) normalisations has been provided by Mullineaux et al. (2004). A number of key features emerged from these normalisations:

- Choosing a defined measure external to the data was recommended. Although this does not reduce the variability the most, defined measures within the data result in 'local' variability reductions. For instance, using the initial value resulted in the variability reducing to zero at the beginning of the data, which was the instant the foot made ground contact.
- The complex offset of a multiplicative scatter correction reduced the variability the most. However, it was not recommended, as the shape of each participant's trace was altered which distorted within-individual data comparisons.
- Offset normalisation reduced the variability sufficiently to increase power and result in more variables being significant. Care was recommended in the interpretation of the results as mean angle changes pre- and

post-intervention of only 1° were found statistically significantly different, which was considered lower than the error limit of the data collection.

- Normalisation improved the normality distribution assumption of the data in their study, although it was recommended that following any data transformation the normality distribution should be checked.
- Report variability was recommended for both sets of data, i.e. pre- and post-normalisation, to enable comparisons between studies.

Analysing multiple trials

Two principal methods of reducing variability using ratio and offset normalisations have now been covered. An additional method of reducing variability, becoming more common in biomechanics, is to collect multiple trials and to use the average score. Issues on this have been covered in detail elsewhere, which include that there is a statistical rationale for using the 'average' instead of the 'best' trial (e.g. Kroll, 1957) and, as a simple recommendation, that at least three trials should be used as a compromise between statistical issues and practicality (see Mullineaux et al., 2001). This overcomes the limitation that one trial may not be representative of the performer's 'normal' technique (Bates et al., 1992).

As averaging trials is arguably problematic, as it is in essence creating a mythical trial, it is possible to analyse these multiple trials in a nested ANOVA. In many software packages these analyses are not straightforward to perform or analyse, and, more importantly, it is unclear how many trials should be used, which affects the statistical power – hence recommendations on its use still need to be investigated. A further approach is to use single-subject analyses where the multiple trials by a single participant provide the individual measures in statistical analyses. Generally, between-group statistics are used (Bates, 1996) as otherwise trial order influences the statistical significance value obtained. Nevertheless if trial order is relevant or justified (e.g. matching via trial order or ranking trials from worst to best), statistical power can be increased by using a within-group statistic. Single-subject analyses overcome another potential limitation of group analyses that, if similarities in performers' techniques do not exist, then the data may represent a mythical 'average' performer (Dufek et al., 1995). Where similarities do not exist, then this decreases statistical power. Examples of the application of single-subject analyses include Hreljac (1998) and Wheat et al. (2003). These analyses can also provide 'baseline' variability for a participant that can be analysed in longitudinal studies. The use of multiple trials also paves the way for analysing variability itself, which is covered in the next section.

ANALYSING VARIABILITY IN TIME-SERIES DATA

Rather than reducing variability for statistical purposes, it can be meaningful to analyse and discuss variability in light of its importance in the successful control and outcomes of movement. In analysing variability, i.e. the degree of departure from the central score, greater measures have been viewed as indicative of poor technique (Davids et al., 1997), functional in producing the desired outcome (Arutyunyan et al., 1968) and reducing injury (Hamill et al., 1999; Heiderscheit et al., 1999; 2000). Conversely, for example, less variability has been viewed as an indication of injury (Peham et al., 2001). In particular, variability can provide a measure of coordination, i.e. the functional link between the muscles and joints used to produce the desired performance or outcome. A variety of methods exist to quantify variability.

With respect to kinematic data, which provide the researcher with a simplified representation of human movement, often discrete values are reported for single variables or combinations of several variables (e.g. ratio of hip to knee angles). These discrete values may be for key instances (e.g. start of movement), a descriptive statistic (e.g. minimum, maximum) and representative of the entire movement (e.g. mean). There are several methods available for quantifying variability within these discrete values. Five methods, namely standard deviation (SD), root mean square difference (RMSD), 95 per cent confidence intervals (95%CI), percentage coefficient of variation (%CV) and percentage RMSD (%RMSD), have been considered previously (Mullineaux et al., 2001). Assuming that RMSD is calculated based on the mean score providing the criterion value (although independent criterion scores can be used such as in an error analysis) then, if the mean, SD and *n* are provided, all of these methods are determinable from each other, facilitating comparisons between different measures used in the literature. In general, if $n>3$, then 95%CI provides the smallest measure followed by RMSD and then SD.

The analysis of discrete values provides an easily obtainable measure of a performance. However, the reporting of discrete values alone has been criticised, as these fail to capture the dynamic nature of movement (e.g. Baumann, 1992). The dynamic nature of movement can qualitatively be assessed from graphical plots of single or multiple variables. Three principal graphical representations exist: variable-time graphs, variable-variable graphs (e.g. angle–angle plots) and phase-plane portraits (see Chapter 3 also). If multiple trials are plotted then, in addition to qualitatively assessing the patterns of movement, variability can be assessed. As human movement is complex, then analysing variability using multiple variables may provide the most comprehensive information (Mullineaux and Wheat, 2002). Multiple variables are, or can be, presented in all three of these graphical plots and used to assess variability over the entire kinematic trace, provided that there are two or more trials plotted. It is recommended, however, that more trials are used. For instance, in calculating the confidence intervals of bivariate plots using CI2 it is recommended that a minimum of ten trials are used (Mullineaux, 2017), which provides a sufficiently low bias between the ellipse area and actual area (Jackson et al., 2011).

Each of these plots is used in a different manner. To illustrate these plots, the knee and hip angles for three trials of a non-injured, male participant running at 3 m s^{-1} are used in all Figures 10.1 to 10.8. With many of these techniques it is necessary that the trials are of the same length. Consequently, these trials were normalised to 101 data points, which is also a beneficial length as it

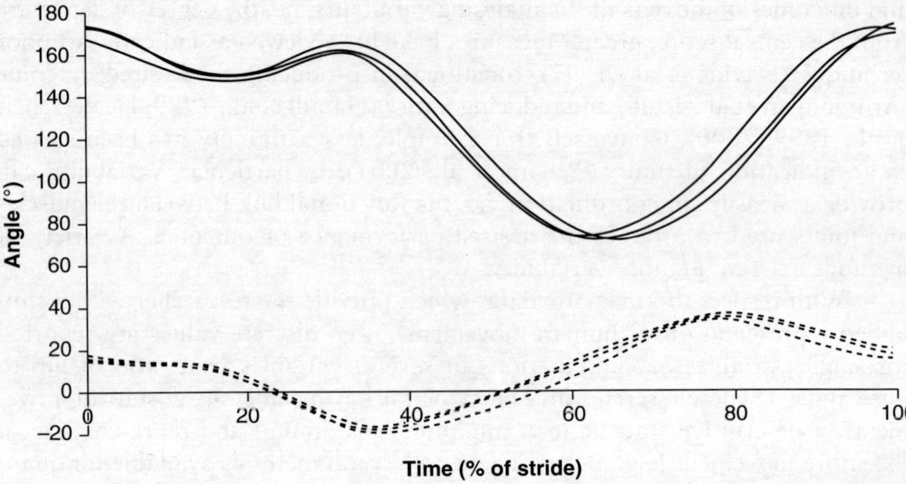

Figure 10.1 Angles for knee (solid lines) and hip (dashed lines) for three trials of a healthy male participant running at 3 m s⁻¹. In the anatomical standing position, the knee is at 180° (flexion positive) and the hip is at 0° (thigh segment to the vertical; flexion positive; hyper-extension negative). Key instances are right foot contacts at 0 per cent and 100 per cent, and right foot off at 40 per cent.

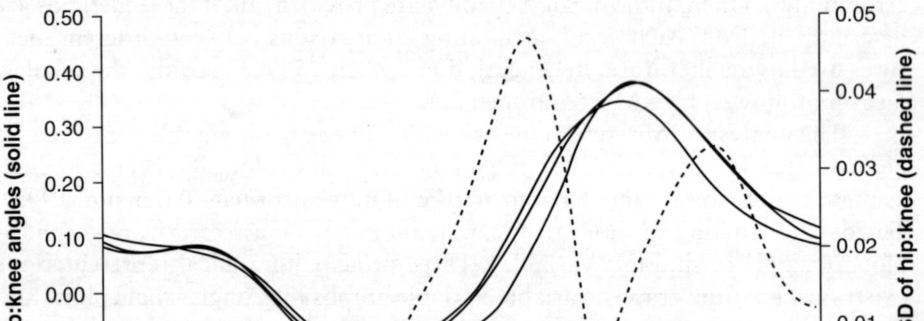

Figure 10.2 Ratio of the hip to the knee angles for three trials of a healthy male participant running at 3 m s⁻¹ (left axis), and using the mean score as the criterion the RMSD of these three trials (right axis).

facilitates discussing percentage of time throughout the movement and it is easy to scale the *x*-axis (i.e. scale time from 0 to 100 per cent of movement).

First, the variable-time graph of the knee and hip angles for three trials are presented in Figure 10.1. In the anatomical standing position, the knee is at 180° (flexion positive) and the hip is at 0° (thigh segment to the vertical; flexion positive; hyper-extension negative). These two variables have been combined as a ratio of the hip to the knee angles (see Figure 10.2; left axis). The ratio

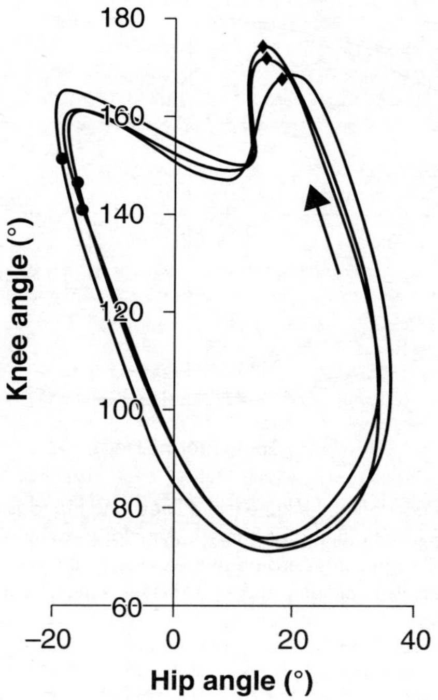

Figure 10.3 Knee-hip angle–angle diagram for three trials of a healthy male participant running at 3 m s^{-1}. Heel strike (♦), toe off (•) and direction (arrow) indicated.

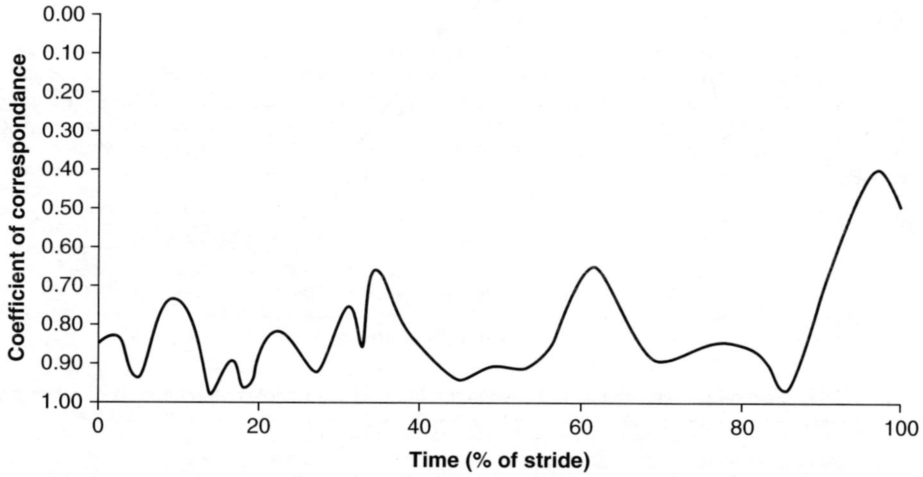

Figure 10.4 Coefficient of correspondence (*r*) determined using vector coding of three trials of the knee-hip angle–angle data for a healthy male participant running at 3 m s^{-1}. The coefficient ranges from maximal variability (*r*=0) to no variability (*r*=1).

of the two variables at each instant over the trials can be used to calculate the variability in the data. For instance, using the RMSD, with the mean as the criterion value, the variability at each instance in the running cycle is presented in Figure 10.2 (right axis).

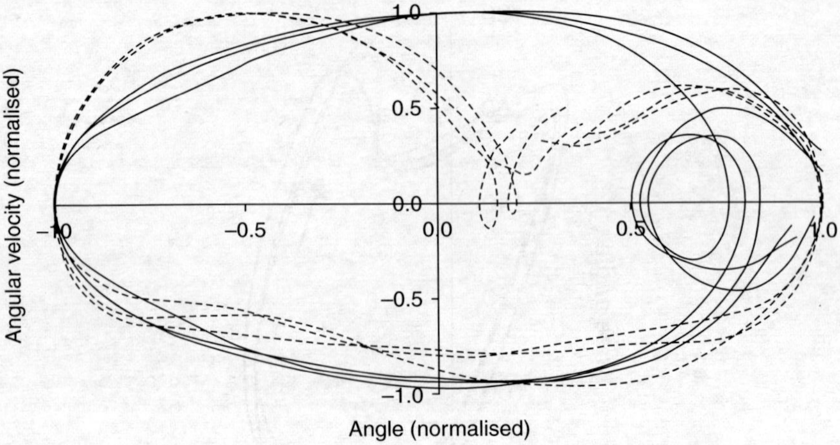

Figure 10.5 Phase-plane portrait of the knee (solid lines) and hip (dashed lines) angles for three trials of a healthy male participant running at 3 m s^{-1}. Angular velocity is normalised to the maximum value across trials (hence 0 represents zero angular velocity), and angle is normalised to the range within trials (i.e. – 1 represents minimum, and +1 represents maximum value).

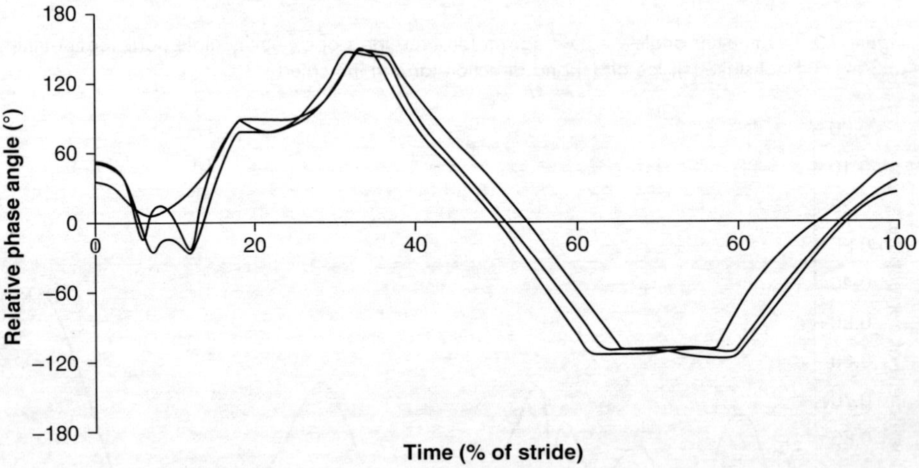

Figure10.6 Continuous relative phase between the hip and knee angles of three trials of a healthy male participant running at 3 m s^{-1}. Phase-plane angle (φ) used in the range of 0°≤ φ ≤ 180°.

Second, the angle–angle plot of the knee-hip is presented in Figure 10.3. To quantify the variability in these data, vector coding can be applied (Tepavac and Field-Fote, 2001). There are two main approaches to vector coding that incorporate both the vector angles and magnitudes (Tepavac and Field-Fote, 2001) or include just the vector angles (Heiderscheit et al., 2002). Vector coding may be applicable to data that are not necessarily linearly related (as is assumed in cross correlations – see later). Based on the method incorporating vector angles

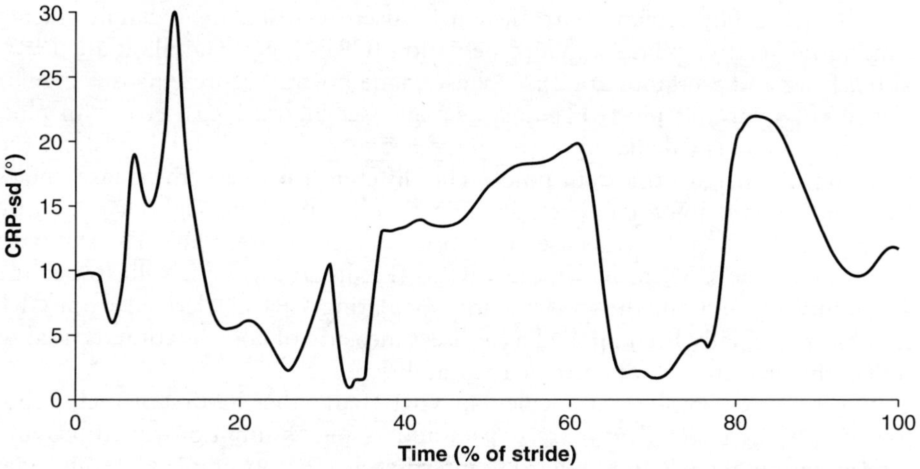

Figure 10.7 Continuous relative phase standard deviation (CRP-sd) in the three CRP angles between the hip and knee angles for three trials of a healthy male participant running at 3 m s^{-1}.

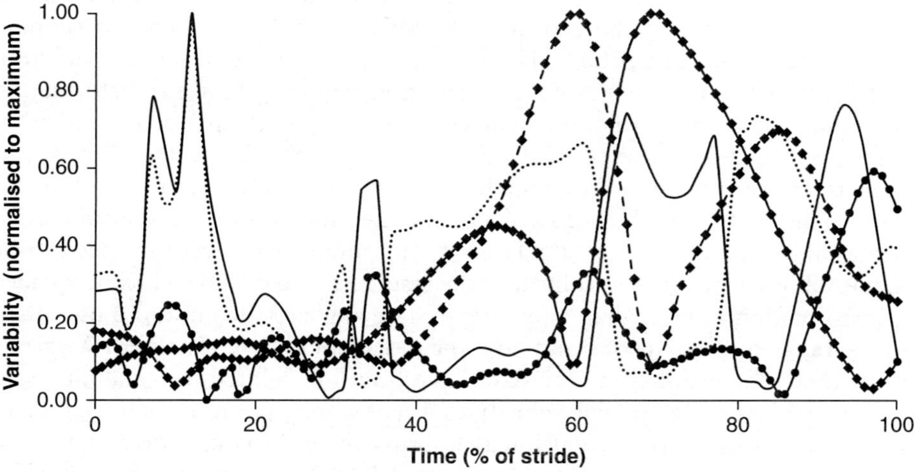

Figure 10.8 Quantification of variability in hip and knee angles for three trials of a healthy male participant running at 3 m s^{-1} using vector coding (•), RMSD (⧫) and continuous relative phase standard deviation (no symbol) for, when in the anatomical standing position, the hip is 0° (solid lines) and hip is 180° (dashed lines). Note, vector coding does not change with the hip angle definition.

and directions, to obtain the vector coding: (1) calculate the vector between each consecutive data point to obtain an angle and magnitude, and repeat this for each trial; (2) at each time instant, determine the mean angle and mean magnitude across the trials, and; (3) obtain the product of the means of the angles and magnitudes to provide the coefficient of correspondence (r). Together, this coefficient represents a summary of the angle (shape) and magnitude (size) of the trials, the plot of which indicates the variability throughout the movement (see Figure 10.4). This varies from r=0 (maximum variation) to r=1 (no variability).

Third, a phase-plane portrait can be used to calculate variability using continuous relative phase standard deviation (CRPsd, e.g. Hamill et al., 1999; 2000; Kurz and Stergiou, 2002). A phase-plane portrait represents the angular velocity of a variable plotted against its angle (see Figure 10.5). From this plot, the phase angle is calculated as the angle between the horizontal axis and the line from the origin to a data point. The difference between the phase angles between the two joints at corresponding time intervals throughout the entire cycle provides the relative phase angle or CRP (see Figure 10.6). The standard deviation in the CRP provides the CRPsd (Figure 10.7), which illustrates the variability throughout the movement. Calculating CRPsd, derived from CRP that has principally been applied as a direct measure of coordination, is used to indirectly measure coordination via variability.

It is more complicated to calculate CRP than either RMSD or vector coding, which has resulted in more papers addressing assumptions, methods and interpretation of CRP. With regard to interpreting CRP, in simple and sinusoidal movements, it is considered to provide a good indication of two joints being in-phase (CRP=0°) or out-of-phase (CRP=180°; Hamill et al., 2000). For more complex motions, such as in running, despite qualitatively similar phase-plane portraits for the hip and knee, it was concluded that CRP provided little informative information on the coordination between these two joints in running (Mullineaux and Wheat, 2002). This was principally attributable to the violation of the assumption of a sinusoidal distribution requiring that a transformation be applied. One transformation includes normalising to the period of oscillation (Peters et al., 2003). These authors also propose sophisticated transforms (e.g. Hilbert Transform), non-linear methodologies and interpolation of positional data as alternative methods to account for the violation of the sinusoidal distribution. Consequently, in calculating CRP a number of decisions need to be made, all of which will radically alter the results. These include which frequency transformation to use, what component phase angle (φ) definition to use (often in the range of $0° \le \varphi \le 180°$), and how the data should be normalised. A variety of methods for normalising the data have been highlighted (see Hamill et al., 2000; e.g. normalise angular velocity to 1 as the maximum of each trial, or to the maximum of all trials), which may improve the one to one ratio assumption. In contrast, as the arc tangent is used to calculate the component phase angle it has been considered that this will account for differences in amplitudes between segments and, as such, no normalising is required (Kurz and Stergiou, 2002).

In directly measuring coordination it is possible to base this on only one trial, whereas measuring coordination indirectly through variability requires two or more trials. In addition to CRP as a direct measure of coordination, cross correlation determined from angle–angle plots is also available. Cross-correlations provide a linear correlation coefficient between two sets of time-series data where one variable has a time lag introduced (Amblard et al., 1994). A summary of recommendations in the use of cross-correlations have been made previously (Mullineaux et al., 2001), and include that as high correlation coefficients can occur for non-linear data, the linearity assumption across the entire data range should be checked visually (i.e. plotting one variable with no time lag against the other variable with its time lag introduced). Previously it has been demonstrated that the relationship between the hip and knee in running was linear in parts but non-linear at other times (Mullineaux and Wheat, 2002). Potentially, analysing the data in phases as has been performed previously (Hamill et al., 1999) may be possible, and applying a linear transformation to the non-linear phase of the movement may improve the cross correlations data analysis.

The techniques mentioned so far (listed in Table 10.3) have principally been described for calculating a measure at each instant throughout the movement cycle. Many of these methods can also be used, by taking an average of these measures across the entire cycle, to provide an overall measure of the variability across the entire movement. In addition to these techniques, another measure simply provides this overall measure of variability – based on an angle–angle plot, NoRMS can be calculated, a technique which has been covered in depth elsewhere (see Sidaway et al., 1995; Mullineaux et al., 2001).

All techniques mentioned so far in this section are useful for analysing variability in single, or pairwise coordination of, variables. When motions involving multiple degrees of freedom – as with most sport and exercise movements – are analysed, multiple single variable or multiple pairwise analyses of coordination can be performed. However, these approaches not only inflate the familywise error rate of any inferential statistics but they fail to account for potential covariance between degrees of freedom and do not provide information on the coordination and variability of the task as a whole (Forner-Cordero et al., 2007); Daffertshofer et al. (2004) suggested that multivariate analysis techniques such as Principal Components Analysis (PCA) could be useful for analysing complex coordination patterns. PCA identifies mutual information in data by considering the covariance between variables. The technique reduces the dimensionality of the dataset by identifying a smaller number of uncorrelated variables known as Principal Components (PC). In n-dimensional space (where n is the number of variables in a dataset), PCA successively identifies the orientation of axes (PCs) that explain the greatest variation in the dataset – with the variation associated with the preceding PC removed before the next PC is identified. This ensures that all PCs are orthogonal in the n-dimensional space, enabling the conversion of the dataset into a set of independent, uncorrelated variables. Most human movements can be captured with only a few principal components, drastically reducing the dimensionality of the data.

Often in biomechanics, time series of continuous data are recorded and analysed. Although, traditionally, PCA has been used to analyse discrete data, the technique has been applied to continuous waveforms and to assess inter-joint coordination. For example, Forner-Cordero et al. (2007) used PCA to examine the complex interaction between multi-joint coordinative movements and postural control. An alternative approach to analysing continuous waveforms is functional Principal Components Analysis (fPCA: Donà et al., 2009; Preatoni et al., 2013). With fPCA all data waveforms are considered as functions rather than long sequences of data, with the resulting principal components represented by functions rather than vectors. An assumption is made that continuous waveforms are best captured by functions rather than strings of numbers and that the data are associated with some underlying, governing functional relationship (Donà et al., 2009). The technique has been used, for example, to analyse the effects of orthotics on lower extremity kinematics in participants with

Table 10.3 Statistical analyses available for quantifying variability (and consequently coordination) in two or more trials, across the entire cycle or as an overall measure for the entire cycle. The examples relate to three trials of a healthy male participant running at 3 m s^{-1} (see Figures 10.2 to 10.7).

Basis	*Technique (recommended reading)*	*Example output*[a]	
		Entire cycle	*Overall*
Variable-time graph (e.g. Fig. 10.1; Fig. 10.2, left axis)	95%CI, SD, RMSD, %CV and %RMSD of a variable or ratio of variables (Mullineaux et al., 2001)	Fig. 10.2 (right axis; RMSD of a ratio of 2 variables)	0.03
Angle–angle plot[b] (e.g. Fig. 10.3)	Vector coding (Tepavac and Field-Fote, 2001)	Fig. 10.4	0.81[c]
	NoRMS (Sidaway et al., 1995; Mullineaux et al., 2001)	N/A	4.2%
	Cross correlation, *r* (Amblard et al., 1994; Mullineaux et al., 2001)	N/A	0.83[d]
	Cross correlation, time lag (Amblard et al., 1994; Mullineaux et al., 2001)	N/A	–18%[d]
Phase plane portrait[b] (e.g. Fig. 10.5)	Continuous Relative Phase, CRP (Hamill et al., 2000)[e]	Fig. 10.6	N/A
	CRP standard deviation, CRPsd (Hamill et al., 2000)	Fig. 10.7	7.6°

[a]N/A indicates output not available from the technique.
[b]Confidence intervals of bivariate plots can be calculated using CI2 (Mullineaux, 2017) but, as a minimum of ten trials is recommended, this is not presented.
[c]Data are a product of magnitude and angle components, both of which can be analysed independently, including just the angles using the method by Heiderscheit et al. (2002).
[d]Average presented of the three trials for *r* (0.827; 0.830; 0.784) and the time lag (–17%; –18%; –17%; negative value indicates the hip moves before the knee).
[e]Provides a phase angle, which is a measure of coordination and not variability.

and without a history of Achilles tendon injury (Donoghue et al., 2008). fPCA has the advantage that it does not make the assumption that the data points of the original curve are independent (Preatoni et al., 2013). Also, time normalisation of data to a consistent length is not required and the technique incorporates smoothing and calculation of derivatives – which, if done separately from the analysis procedure can result in unwanted and unknown sources of variation in the data (Preatoni et al., 2013).

Other contemporary statistical analysis techniques offer the possibility to perform inferential statistical tests on multiple time series of continuous data. Statistical Parametric Mapping (SPM) – which was originally developed for the analysis of cerebral blood flow with functional magnetic resonance imaging – provides a framework for the analysis of smooth bounded n-dimensional fields (Pataky, 2010). The technique has recently been applied in a biomechanics context to analyse common '1D trajectories', such as joint angles and electromyography signals (Pataky et al., 2013; Robinson et al., 2015) as well as 2D pressure and 3D strain fields (Pataky, 2010).

Commonly in biomechanics, differences between experimental conditions are assessed by extracting single, scalar values from the time series of continuous waveforms (e.g. maximum or minimum values; value at a specific time point) for comparison. Pataky and colleagues (Pataky, 2010; Pataky et al., 2013, 2014; Robinson et al., 2015) argue that this introduces two potential sources of bias for some hypothesis tests: (1) 'regional focus bias' – where focusing on one region of time in the continuous waveform data introduces bias, and (2) 'inter-component covariation bias' – where there is a failure to consider the covariation between the different continuous waveforms included in an analysis (e.g. multiple joint angles in a kinematic analysis). SPM is conceptually similar to more traditional inferential statistics but instead of using traditional probability distributions to assess the statistical significance of differences it addresses these sources of bias using probability distributions from Random Field Theory (Pataky et al., 2013). By considering continuous waveform data as scalar fields (rather than extracting single scalar values from the time series), regional focus bias in non-directed hypothesis tests can be addressed. Further, by considering multiple scalar fields in one vector field, all continuous waveforms are combined into a sole unit of observation (Robinson et al., 2015), addressing the inter-component covariance bias. Importantly, when analysing multiple continuous waveforms with potential for covariance, it is possible that no significant differences between conditions would emerge across the data set in its entirety (the vector field) at the same time as differences between certain individual waveforms (the scalar fields within the vector field) being statistically significant. Conversely, it is also possible that the differences for all individual waveforms are not statistically significant but the differences for the entire vector field are significant.

In summary, analysis of variability can be performed using a variety of techniques. These tend to provide different 'qualitative' findings, as the periods of small and large variability that may be used to infer any coordination strategies vary between the techniques (see Figure 10.8; solid lines). In addition, quantitatively the results are also difficult to compare (see Table 10.3). The choice of which to use may be dependent on meeting underpinning assumptions, ease of use and interpretation, ability to compare to previous literature and preference of the researcher. Some comparisons in the literature include that, for example, cross correlations were considered to provide a more meaningful measure of coordination than CRP (Mullineaux and Wheat, 2002). It was considered that a transformation technique may improve the assumptions for CRP, but that the greater data processing may not be warranted as it could make the calculation and interpretation too complicated. A further consideration is that only vector coding (compared to RMSD and CRPsd) was found to provide the same results when the angle definition of the hip was altered (i.e. full extension changed from 0° to 180°), as demonstrated in Figure 10.8 (compare solid to dashed lines). Consequently, as vector coding is the only reliable measure (i.e. not dependent on the variable definition) and it is simple to calculate, it may prove to be the most suitable technique for quantifying variability, but other techniques may become more viable through further development or standardisation. When analysing motions involving multiple degrees of freedom, insights from statistical analysis with SPM could have an effect on the interpretation of results in biomechanical studies of complex, coordinated human motion. SPM, together with other contemporary techniques such as PCA and fPCA, are more challenging to implement but offer potentially insightful and relevant biomechanical analysis into complex, multiple degrees of freedom human movement systems.

REPORTING OF A STUDY

Reporting research in an effective manner facilitates the dissemination of knowledge and subsequent advancement of science. A key feature in reporting research is to work within the pressure on publication space. In this chapter, which focused on a small area of research methods, many factors were highlighted that can affect results and consequently the meaning of data. For instance, angle definitions alter measures of variability and the method of treating trials affects statistical power. But, for example, in 20 per cent of a selection of biomechanical papers the method of treating trials was unclear (Mullineaux et al., 2001). Nevertheless, in combination with the various chapters of this book, it is suggested that you aim to:

- Report sufficient details to enable the reader to understand the theoretical underpinning of the research question, use the findings to support future research, replicate the study and interpret the findings.
- Cite references to standard procedures and statistical analyses (e.g. see references cited throughout this chapter).
- Provide details on the defined population, sampling frame and sampling method, and report the sample characteristics.
- Outline the research design, including the timing, number and sequence of measurements, and provide details on trial sizes used and method of treating trials.
- Clarify methods to control for experimental errors (e.g. calibration procedures).
- Address the reliability of the measurements.
- Provide informative results, particularly descriptive statistics as they provide simple data of practical value. Also reporting these in the abstract for key variables is beneficial as it allows the reader to perform a simple power analysis.
- Justify the level of inferential statistical significance (α) and probability values (P), and report an *a-priori* statistical power analysis and exact P values for results.
- Justify the statistical analyses used.
- Report assessments and corrections applied to the assumptions of statistical tests.
- Use effect size statistics to supplement, or supplant, significance tests.
- Establish and quantify uncertainties or errors in the data.
- Present information in the most appropriate format either in a chart, if a visual inspection is required, or a table, if exact values are important, but duplication should be avoided.
- Focus the report on the research question that was addressed, the underlying theory, previous related research, the importance of the results of the study and how they contribute to existing knowledge, and assess the limitations of the study.
- List references accurately.

REFERENCES

Amblard, B., Assaiante, C., Lekhel, H. and Marchand, A. R. (1994) 'A statistical approach to sensorimotor strategies: conjugate cross-correlations', *Journal of Motor Behavior*, 26: 103–112.

Arribas-Gil, A. and Romo, J. (2014) 'Shape outlier detection and visualization for functional data: the outliergram', *Biostatistics*, 15: 603–619.

Arutyunyan, G. H., Gurfinkel, V. S. and Mirskii, M. L. (1968) Investigation of aiming at a target, *Biophysics*, 13: 536–538.

Back, W., Barneveld, A., Bruin, G., Schamhardt, H. C. and Hartman, W. (1994) 'Kinematic detection of superior gait quality in young trotting Warmbloods', *The Veterinary Quarterly*, 16S2: 91–96.

Bates, B. T. (1996) 'Single subject methodology: an alternative approach', *Medicine and Science in Sports and Exercise*, 28: 631–638.

Bates, B. T., Dufek, J. S. and Davis, H. P. (1992) 'The effect of trial size on statistical power', *Medicine and Science in Sports and Exercise*, 24: 1059–1068.

Batterham, A. M., George, K. P. and Mullineaux, D. R. (1997) 'Allometric scaling of left ventricular mass and body dimensions in males and females', *Medicine and Science in Sports and Exercise*, 29: 181–186.

Baumann, W. (1992) 'Perspectives in methodology in biomechanics of sport' in R. Rodano, G. Ferrigno and G. Santambrogio (eds), *Proceedings of the Xth Symposium of the International Society of Biomechanics in Sports*, pp. 97–104, Milan, Italy: Edi Ermes.

Baumgartner, T. A. and Strong, C. H. (1998) *Conducting and Reading Research in Health and Human Performance*. Boston, MA: McGraw-Hill.

Burden, A. M., Grimshaw, P. N. and Wallace, E. S. (1998) 'Hip and shoulder rotations during the golf swing of sub-10 handicap players', *Journal of Sports Sciences*, 16: 165–176.

Carver, R. P. (1978) 'The case against statistical significance testing', *Harvard Educational Review*, 48: 378–399.

Cham, R. and Redfern, M. S. (2002) 'Changes in gait when anticipating slippery floors', *Gait and Posture*, 15: 159–171.

Cohen, J. (1988) *Statistical Power Analysis for the Behavioural Sciences*, 2nd edn. Hillsdale, NJ: Lawrence Erlbaum.

Cohen, J. (1992) 'A power primer', *Psychological Bulletin*, 112: 155–159.

Daffertshofer, A., Lamoth, C. J. C., Meijer, O. G. and Beek, P. J. (2004) 'PCA in studying coordination and variability: a tutorial', *Clinical Biomechanics*, 19: 415–428.

Davids, K., Bennett, S. J., Handford, C. H., Jolley, L. and Beak, S. (1997) 'Acquiring coordination in interceptive actions: an ecological approach', in R. Lidor and M. Bar-Eli (eds), *Innovations in Sport Psychology: Linking Theory and Practice*, pp. 227–229, Netanya, Israel: International Society of Sport Psychology.

Degueurce, C., Pourcelot, P., Audigié, F., Denoix, J. M. and Geiger, D. (1997) 'Variability of the limb joint patterns of sound horses at trot', *Equine Veterinary Journal*, 23S: 89–92.

Donà, G., Preatoni, E., Cobelli, C., Rodano, R. and Harrison, A. J. (2009) 'Application of functional principal component analysis in race walking: an emerging methodology', *Sports Biomechanics*, 8: 284–301.

Donoghue, O. A, Harrison, A. J., Coffey, N. and Hayes, K. (2008) 'Functional data analysis of running kinematics in chronic Achilles tendon injury', *Medicine and Science in Sports and Exercise*, 40: 1323–1335.

Dufek, J. S., Bates, B. T., James, C. J. and Stergiou, N. (1995) 'Interactive effects between group and single-subject response patterns', *Human Movement Science*, 14: 301–323.

Duffey, M. J., Martin, D. F., Cannon, D. W., Craven, T. and Messier, S. P. (2000) 'Etiologic factors associated with anterior knee pain in distance runners', *Medicine and Science in Sports and Exercise*, 32: 1825–1832.

Duncan, T. (1987) *Physics: A Textbook for Advanced Level Students*. London: John Murray.

Forner-Cordero, A., Levin, O., Li, Y. and Swinnen, S. P. (2007) 'Posture control and complex arm coordination: analysis of multijoint coordinative movements and stability of stance', *Journal of Motor Behavior*, 39: 215–226.

Friendly, M. (2006) Power analysis for ANOVA designs. www.math.yorku.ca/SCS/Online/power/, visited (accessed 19 January 2017).

Hamill, J., Haddad, M. and McDermott, W. J. (2000) 'Issues in quantifying variability from a dynamical systems perspective', *Journal of Applied Biomechanics*, 16: 407–418.

Hamill, J., van Emmerik, R. E. A., Heiderscheit, B. C. and Li, L. (1999) 'A dynamical systems approach to lower extremity running injuries', *Clinical Biomechanics*, 14: 297–308.

Heiderscheit, B. C. (2000) 'Movement variability as a clinical measure for locomotion', *Journal of Applied Biomechanics*, 16: 419–427.

Heiderscheit, B. C., Hamill, J. and van Emmerik, R. E. A. (1999) 'Q angle influences on the variability of lower extremity coordination', *Medicine and Science in Sports and Exercise*, 31: 1313–1319.

Heiderscheit, B. C., Hamill, J. and van Emmerik, R. E. A. (2002) 'Variability of stride characteristics and joint coordination among individuals with unilateral patellofemoral pain', *Journal of Applied Biomechanics*, 18: 110–121.

Holmström, M., Fredricson, I. and Drevemo, S. (1994) 'Biokinematic differences between riding horses judged as good and poor at the trot', *Equine Veterinary Journal*, 17S: 51–56.

Hopkins, W. G., Hawley, J. A. and Burke, L. M. (1999) 'Design and analysis of research on sport performance enhancement', *Medicine and Science in Sports and Exercise*, 31: 472–485.

Howell, D. C. (1997) *Statistical Methods for Psychology*, Belmont, CA: Duxbury Press.

Hreljac, A. (1998) 'Individual effects on biomechanical variables during landing in tennis shoes with varying midsole density', *Journal of Sports Sciences*, 16: 531–537.

Jackson, A. L., Inger, R., Parnell, A. C. and Bearhop, S. (2011) 'Comparing isotopic niche widths among and within communities: SIBER - Stable Isotope Bayesian Ellipses in R', *Journal of Animal Ecology*, 80: 595–602.

Jaric, S. (2002) 'Muscle strength testing: use of normalisation for body size', *Sports Medicine*, 32: 615–631.

Krejcie, R. V. and Margan, D. W. (1970) 'Determining sample size for research activities', *Education and Psychological Measurement*, 30: 607–610.

Kroll, W. (1957) 'Reliability theory and research decision in selection of a criterion score', *Research Quarterly*, 38: 412–419.

Kurz, M. J. and Stergiou, N. (2002) 'Effect of normalization and phase angle calculations on continuous relative phase', *Journal of Biomechanics*, 35: 369–374.

Leys, C., Ley, C., Klein, O., Bernard, P. and Licata, L. (2013) 'Detecting outliers: Do not use standard deviation around the mean, use absolute deviation around the median', *Journal of Experimental and Social Psychology*, 49: 764–766.

Lenth, R. V. (2001) 'Some practical guidelines for effective sample-size determination', *The American Statistician*, 55: 187–193.

Matthews, R. (1998) 'Flukes and flaws', *Prospect*, 35: 20–24.

Mullineaux, D. R. (2017) 'CI2 for creating and comparing confidence-intervals for time-series bivariate plots', *Gait and Posture*, 23: 367–373.
Mullineaux, D. R., Barnes, C. A. and Batterham, A. M. (1999) 'Assessment of bias in comparing measurements: a reliability example', *Measurement in Physical Education and Exercise Science*, 3: 195–205.
Mullineaux, D. R. and Bartlett, R. M. (1997) 'Research methods and statistics', in R. M. Bartlett (ed.), *Biomechanical Analysis of Movement in Sport and Exercise*, pp. 81–104, Leeds: BASES.
Mullineaux, D. R., Bartlett, R. M. and Bennett, S. (2001) 'Research methods and statistics in biomechanics and motor control', *Journal of Sports Sciences*, 19: 739–760.
Mullineaux, D. R., Clayton, H. M. and Gnagey, L. M. (2004) 'Effects of offset-normalizing techniques on variability in motion analysis data', *Journal of Applied Biomechanics*, 20: 177–184.
Mullineaux, D. R. and Irwin, G. (2017) 'Error and anomaly detection for intra-participant time-series data', *International Biomechanics*, 4: 28–35.
Mullineaux, D. R., Milner, C. E., Davis, I. S. and Hamill, J. (2006) 'Normalization of ground reaction forces', *Journal of Applied Biomechanics*, 22: 230–233.
Mullineaux, D. R. and Wheat, J. (2002) 'Quantifying coordination in kinematic data: a running example', in K. E. Gianikellis (ed.), *Proceedings of the 20th International Symposium on Biomechanics in Sport*, pp. 515–518, Cáceres, Spain: University of Extremadura.
Næs, T., Isaksson, T., Fearn, T. and Davies, T. (2002) *A User-friendly Guide to Multivariate Calibration and Classification*, Chichester, England: NIR Publications.
Park, I. and Schutz, R. W. (1999) ' "Quick and easy" formulae for approximating statistical power in repeated measures ANOVA', *Measurement in Physical Education and Exercise Science*, 4: 249–270.
Pataky, T. C. (2010) 'Generalized n-dimensional biomechanical field analysis using statistical parametric mapping', *Journal of Biomechanics*, 43: 1976–1982.
Pataky, T. C., Robinson, M. A. and Vanrenterghem, J. (2013) 'Vector field statistical analysis of kinematic and force trajectories', *Journal of Biomechanics*, 46: 2394–2401.
Pataky, T. C., Robinson, M. A, Vanrenterghem, J., Savage, R., Bates, K. T. and Crompton, R. H. (2014) 'Vector field statistics for objective center-of-pressure trajectory analysis during gait, with evidence of scalar sensitivity to small coordinate system rotations', *Gait and Posture*, 40: 255–258.
Peham, C., Licka, T., Girtler, D. and Scheidl, M. (2001) 'The influence of lameness on equine stride length consistency', *Veterinary Journal*, 162: 153–157.
Peters, B. T., Haddad, J. M., Heiderscheit, B. C., van Emmerik, R. E. and Hamill, J. (2003) 'Limitations in the use and interpretation of continuous relative phase', *Journal of Biomechanics*, 36: 271–274.
Preatoni, E., Hamill, J., Harrison, A. J., Hayes, K., Van Emmerik, R. E. A, Wilson, C. and Rodano, R. (2013) 'Movement variability and skills monitoring in sports', *Sports Biomechanics*, 12: 69–92.
Pullinen, T., Mero, A., Huttunen, P., Pakarinen, A. and Komi, P. V. (2002) 'Resistance exercise induced hormonal responses in men, women, and pubescent boys', *Medicine and Science in Sports and Exercise*, 34: 806–813.
Raftopoulos, D. D., Rabetas, D. A., Armstrong, C. W., Jurs, S. G. and Georgiadis, G. M. (2000) 'Evaluation of an existing and a new technique for the normalization of ground reaction forces: total body weight versus lean body weight', *Clinical Kinesiology*, 4: 90–95.
Robinson, M. A., Vanrenterghem, J. and Pataky, T. C. (2015) 'Statistical parametric mapping (SPM) for alpha-based statistical analyses of multi-muscle EMG time-series', *Journal of Electromyography and Kinesiology*, 25: 14–19.

Sangeux, M. and Polak, J. (2014) 'A simple method to choose the most representative stride and detect outliers', *Gait and Posture*, 41: 726–730.

Schmidt-Nielsen, K. (1984) *Scaling: Why Is Animal Size So Important?* Cambridge: Cambridge University Press.

Shultz, B. B. and Sands, W. A. (1995) Understanding measurement concepts and statistical procedures, in P. J. Maud and C. Foster (eds), *Physiology Assessment of Human Fitness*, pp. 257–287, Champaign, IL: Human Kinetics.

Sidaway, B., Heise, G. and Schonfelder-Zohdi, B. (1995) 'Quantifying the variability of angle-angle plots', *Journal of Human Movement Studies*, 29: 181–197.

Tabachnick, B. G. and Fidell, L. S. (2001) *Using Multivariate Statistics*, Boston, MA: Allyn and Bacon.

Tanner, J. M. (1989) *Foetus into Man*, Ware, England: Castlemead.

Tepavac, D. and Field-Fote, E. C. (2001) 'Vector coding: a technique for quantification of intersegmental coupling behaviours', *Journal of Applied Biomechanics*, 17: 259–270.

Tyler, R. W. (1931) 'What is statistical significance?' *Educational Research Bulletin*, 10: 115–118.

van Weeren, P. R., van den Bogert, A. J. and Barneveld, A. (1992) 'Correction models for factors for skin displacement in equine kinematic gait analysis', *Journal of Equine Veterinary Science*, 12: 178–192.

Vincent, W. V. (1999) *Statistics in Kinesiology*, Champaign, IL: Human Kinetics.

Wheat, J. S., Bartlett, R. M., Milner, C. E. and Mullineaux, D. R. (2003) 'The effect of different surfaces on ground reaction forces during running: a single-individual design approach', *Journal of Human Movement Studies*, 44: 353–364.

Winer, B. J., Brown, D. R. and Michels, K. M. (1991) *Statistical Principles in Experimental Design*, New York: McGraw-Hill.

Winter, E. M. and Nevill, A. M. (2001) 'Scaling: adjusting for differences in body size', in R. Eston and T. Reilly (eds), *Kinanthropometry and Exercise Physiology Laboratory Manual: Tests, Procedures and Data. Volume 1: Anthropometry*, pp. 275–293, London: Routledge.

CHAPTER 11

COMPUTER SIMULATION MODELLING IN SPORT

Maurice R. Yeadon and Mark A. King

INTRODUCTION

Experimental science aims to answer research questions by investigating the relationships between variables using quantitative data obtained in an experiment and assessing the significance of the results statistically (Yeadon and Challis, 1994). In an ideal experiment the effects of changing just one variable are determined. While it may be possible to change just one variable in a carefully controlled laboratory experiment in the natural sciences, this is problematic in the sports sciences in general and in sports biomechanics in particular. If a typical performance in a sport such as high jumping is to be investigated, then any intervention must be minimal lest it make the performance atypical. For example, if the intent is to investigate the effect of run-up speed on the height reached by the mass centre in flight, asking the jumper to use various lengths of run-up might be expected to influence jumping technique minimally if the athlete normally does this in training. In such circumstances the run-up speed may be expected to vary as intended but other aspects of technique may also change. Faster approaches may be associated with a greater stride length and a more horizontal planting of the take-off leg. As a consequence, the effects of a faster approach may be confounded by the effects of larger plant angles and changes in other technique variables. To isolate the relationship between approach speed and jump height, statistical methods of analysis that remove the effects of other variables are needed (e.g. Greig and Yeadon, 2000). For this to be successful there must be a sufficient quantity and range of data to cope with the effects of several variables.

Theoretical approaches to answering a research question typically use a model that gives a simplified representation of the physical system under study. The main advantage of such a model is that ideal experiments can be carried out since it is possible to change just one variable. This chapter describes theoretical models used in sports biomechanics, detailing their various components and discussing their strengths and weaknesses.

Models may be used to address the forward dynamics problem and the inverse dynamics problem. In the forward dynamics problem the driving forces are specified and the problem is to determine the resulting motion. In the inverse dynamics problem the motion is specified and the problem is to determine the driving forces that produced the motion (Zatsiorsky, 2002). Both of these types of problem are addressed using various modelling approaches, and their relative advantages are discussed.

This chapter first describes the process of building a mathematical model using rigid bodies and elastic structures to represent body segments and various ways of representing the force-generating capabilities of muscle. Direct and indirect methods of determining the physical parameters associated with these elements are described. Before using a model to answer a research question it is first necessary to establish that the model is an adequate representation of the real physical system. This process of model evaluation, by comparing model output with real data, is discussed. Examples of applications of computer modelling are given along with guidelines on conducting a study and reporting it.

THE FORWARD DYNAMICS PROBLEM

In the forward dynamics problem, the driving forces are specified and the problem is to determine the resulting motion. Muscle forces or joint torques may be used as the drivers, in which case the joint angle time histories will be part of the resulting motion. If joint angle time histories are used as drivers for the model then the resulting motion will be specified by the whole body mass centre movement and whole body orientation time history. When a model is used in this way it is known as a simulation model.

Model building

The human body is very complex, with over 200 bones and 500 muscles and, therefore, any human body model will be a simplification of reality. The degree of simplification of a simulation model will depend on the activity being simulated and the purpose of the study. For example a one-segment model of the human body may adequately represent the aerial phase of a straight dive but a model with two or three segments would be required for a piked dive to give an adequate representation. As a consequence a single model cannot be used to simulate all activities and so specific simulation models are built for particular tasks. Furthermore, the complexity of the model can influence the results obtained and conclusions drawn (Mills et al., 2008). As a general rule the model should be as simple as possible, while being sufficiently complex to address the questions set. This simple rule of thumb can be quite difficult to implement since the level of complexity needed is not always obvious.

Essentially, forward dynamics simulation models can either be angle driven, where the joint angle-time histories are input to the model and the resulting whole body orientation and mass centre position are calculated along with the required joint torques, or torque/force driven, where the joint torque or muscle force-time histories are input to the model and the resulting kinematics are calculated. Angle-driven simulation models have typically been used to simulate activities that are not limited by strength, such as the aerial phase of sports movements including diving (Miller, 1971), high jumping (Dapena, 1981) and trampolining (Yeadon et al., 1990a). They have also been used in other activities, such as high bar circling (Yeadon and Hiley, 2000) and long swings on rings (Brewin et al., 2000), by limiting the joint torques to avoid unrealistic movements. Most force-driven or torque-driven simulation models have been used to represent relatively simple planar jumping movements, in which the human body can be represented using simplified planar two-dimensional models. In addition, movements where the body remains symmetrical about the sagittal plane, such as swinging on rings (Sprigings et al., 1998), have often been modelled as this allows the simulation model to have fewer segments and hence fewer degrees of freedom.

Angle-driven models have typically been more complex, with more segments and degrees of freedom, as they are easier to control, while torque-driven models have been relatively simple in general, owing to the difficulties in determining realistic parameters for muscles. One notable exception is the jumping model of Hatze (1981a) which simulated the take-off phase in long jumping. This model was composed of 17 segments and 46 muscle groups, but did not simulate the impact phase and did not allow for soft tissue movement.

Model components

The following sections discuss the various components that are used to build a typical simulation model.

Linked segment models

Most of the whole body simulation models in sports biomechanics are based on a collection of rigid bodies (segments) linked together, and are generically called 'linked segment systems'. The rigid bodies are the principal building blocks of simulation models and can be thought of as representing the basic structure and inertia of the human body. For each rigid segment in a planar model, four parameters are usually required: length, mass, mass centre location and moment of inertia. The number of segments used depends on the aim of the study and the activity being modelled. For example, Alexander (1990) used a two-segment model to determine optimum approach speeds in jumps for height and distance, Neptune and Kautz (2000) used a planar two-legged bicycle-rider model to look at muscle contributions in forward and backward pedalling, and King and Yeadon (2003) used a planar five-segment model to investigate take-offs in tumbling. The complexity needed is not always obvious.

Torque-free two-segment models of vaulting have been used to show that the backward rotation generated during the take-off of a Hecht vault is largely a function of the velocity and configuration at initial contact together with the passive mechanics during impact (Sprigings and Yeadon, 1997; King et al., 1999). These results were confirmed using a torque-driven five-segment model but it was also shown that the inclusion of a hand segment and shoulder elasticity made substantial contributions to rotation (King and Yeadon, 2005).

Wobbling masses

Although linked rigid body models have been used extensively to model many activities, a recent development has been to modify some of the rigid segments in the model by incorporating wobbling mass elements (Gruber et al., 1998; Challis and Pain, 2008). This type of representation allows some of the mass (soft tissue) in a segment to move relative to the bone, the rigid part. For impacts, the inclusion of wobbling masses within the model is crucial as the loading on the system can be up to nearly 50 per cent lower for a wobbling mass model compared with the equivalent rigid segment model (Pain and Challis, 2006). The most common way to model wobbling masses is to attach a second rigid element to the first fixed rigid element, representing the bone, within a segment using non-linear damped passive springs with spring force $F = kx^3 - d\dot{x}$ where x is displacement, $\dot{x}$ is velocity and k and d are constants (Pain and Challis, 2001a).

The disadvantage of including wobbling mass elements within a simulation model is that there are more parameter values to determine and the equations of motion are more complex leading to longer simulation times. Wobbling mass segments should, therefore, only be included when necessary. Whether to include wobbling masses depends on the activity being modelled, although it is not always obvious whether they are needed. For example a simulation model of springboard diving (Yeadon et al., 2006a) included wobbling mass segments, but when the springs were made 500 times stiffer the resulting simulations were almost identical.

Connection between rigid links

Typically the rigid links in the simulation model are joined together by frictionless joints, whereby adjacent segments share a common line or a common point. For example Neptune and Kautz (2000) used a hinge joint to allow for flexion–extension at the knee while Hatze (1981a) used a universal joint at the hip with three degrees of freedom to allow for flexion–extension, abduction–adduction and internal–external rotation. The assumption that adjacent segments share a common point or line is a simplification of reality and, although reasonable for most joints, it is questionable at the shoulder where motion occurs at four different joints. Models of the shoulder joint have ranged in complexity from a one degree of freedom pin joint (Yeadon and King, 2002) to relatively simple viscoelastic representations (Hiley and Yeadon, 2003a) and complex finite element models (van der Helm, 1994). The complexity to be used depends on the requirements of the study. Simple viscoelastic representations have been used successfully in whole body models where the overall movement is of interest, whereas complex models have been used to address issues such as the contribution of individual muscles to movement at the shoulder joint. Furthermore for activities involving impacts, a pin jointed model may not be sufficient for representing internal loading, as force is transmitted through the model too quickly (Allen et al., 2012).

Interface with external surface

The simplest way to model contact between a human body model and an external surface, such as the ground or sports equipment, is to use a 'joint' so that the model rotates about a fixed point on the external surface (Bobbert et al., 2002). The disadvantage of this method is that it does not allow the model to translate relative to the point of contact or allow for a collision with the external surface since, for an impact to occur, the velocity of the point contacting the surface has to be non-zero initially. Alternatively, forces can be applied at a finite number of locations using viscoelastic elements at the interface, with the forces determined by the displacements and velocities of the points in contact. The viscoelastic elements can be used to represent specific elastic structures within the body such as the heel pad (Pain and Challis, 2001b) or sports equipment such as the high bar (Hiley and Yeadon, 2003b), tumble track-foot interface (King and Yeadon, 2004) or landing mat-foot interface (Mills et al., 2006). The equations used for the viscoelastic elements have varied in complexity from simple damped linear representations (King and Yeadon, 2004) through to highly non-linear equations (Wright et al., 1998). The number of points of contact varies but it is typically fewer than three (Yeadon and King, 2002) although 66 points of contact were used to simulate heel–toe running (Wright et al., 1998). The horizontal forces acting while in contact with an external surface can be calculated using a pseudo-Coulomb friction model (Gerritsen et al., 1995) where the horizontal force is expressed as a function of the vertical force and the horizontal velocity of the point in contact or by using viscoelastic springs (Yeadon and King, 2002). If viscoelastic springs are used, the horizontal force should be expressed as a function of the vertical force so that the horizontal force falls to zero at the same time as the vertical force (Wilson et al., 2006). An alternative is to use separate models for static friction (stiction) and sliding friction and switch between them in a simulation (Jackson et al., 2011).

Muscle models

Muscle models in sports biomechanics are typically based upon the work of A.V. Hill where the force-producing capabilities of muscle are divided into contractile and elastic elements (lumped parameter models) with the most commonly used version being the three-component Hill model (Caldwell, 2014).

The model consists of a contractile element and two elastic elements, the series elastic element and the parallel elastic element. Mathematical relationships are required for each element in the muscle model so that the force exerted by a muscle on the simulation model can be defined throughout a simulation.

Contractile element

The force that a contractile element produces can be expressed as a function of three factors; muscle length, muscle velocity and muscle activation. The force-length relationship for a muscle is well documented as being bell-shaped with small tensions at extremes of length and maximal tension in between (Edman, 1992). As a consequence, the force–length relationship is often modelled as a simple quadratic function.

The force–velocity relationship for a muscle can be split into two parts: the concentric phase and the eccentric phase. In the concentric phase, tetanic muscle force decreases hyperbolically with increasing speed of shortening, approaching zero at maximum shortening speed (Hill, 1938). In the eccentric phase, maximum tetanic muscle force increases rapidly to around 1.4–1.5 times the isometric value with increasing speed of lengthening and then plateaus for higher speeds (Dudley et al., 1990; Harry et al., 1990). Maximum voluntary muscle force shows a similar force–velocity relationship in the concentric phase, but plateaus at 1.1–1.2 times the isometric value in the eccentric phase (Westing et al., 1988; Yeadon et al., 2006a).

The voluntary activation level of a muscle ranges from 0 (no activation) to 1 (maximum voluntary activation) during a simulation and is defined as a function of time. This function is multiplied by the maximum voluntary force given by the force–length and force–velocity relationships to give the muscle force exerted. Ideally the function used to define the activation time history of a muscle should have only a few parameters. One way of doing this is to define a simple activation profile for each muscle (basic shape) using a few parameters (Yeadon and King, 2002). For example, in jumping, the activations of the extensors rise up from a low initial value to a maximum and then drop off towards the end of the simulation, while the flexor activations drop from an initial to a low value and then rise towards the end of the simulation (King, Wilson and Yeadon, 2006). These parameters are varied within realistic limits to define the activation-time history used for each muscle during a specific simulation.

Series elastic element

The series elastic element represents the connective tissue – the tendon and aponeurosis – in series with the contractile element. The force produced by the series elastic element is typically expressed as an increasing function of its length with a slack length below which no force can be generated. It is usually assumed that the series elastic element stretches by around 5 per cent at maximum isometric force (Muramatsu et al., 2001).

Parallel elastic element

The effect of the parallel elastic element is often ignored in models of sports movements as this element does not produce high forces for the normal working ranges of joints (Chapman, 1985). Anderson et al. (2007), however, include series elastic effects in their model of joint torque while Begon et al. (2009) apply passive torque considerations in the analysis of hip flexibility in a gymnastics movement on high bar.

Torque generators vs. individual muscle representations

All simulation models that include individual muscle models have the disadvantage that it is very difficult to determine individual parameters for each element of each muscle, as it is impossible to measure all the parameters required non-invasively. As a consequence, researchers rely on data from the literature for their muscle models and so the models are not specific to an individual. An alternative approach is to use torque generators to represent the net effect of all the muscles crossing a joint (e.g. King and Yeadon, 2002) as the net torque produced by a group of muscles can be measured on a constant velocity dynamometer or the extensor and flexor muscle groups around a joint can be represented by separate torque generators (King et al., 2006). More recently the extensor/flexor torque representations about a joint have been separated into mono-articular and bi-articular torque components (King et al., 2012a; Lewis et al., 2012). In all three cases, each torque generator consists of rotational elastic and contractile elements. Using torque generators instead of individual muscles gives similar mathematical relationships with the maximum voluntary torque produced by the contractile element being expressed as a function of the muscle angle and muscle angular velocity (Yeadon et al., 2006b).

Model construction

The following sections will discuss the process of building a simulation model and running simulations using the components described in the previous section.

Free-body diagram of the model

A free-body diagram of a simulation model gives all the necessary information required to build the computer simulation model. The free-body diagram should include the segments, the forces and torques, and the nomenclature for lengths (Figure 11.1). In the system shown in Figure 11.1 there are two degrees of freedom since the two angles θa and θb define the orientation and configuration of the model.

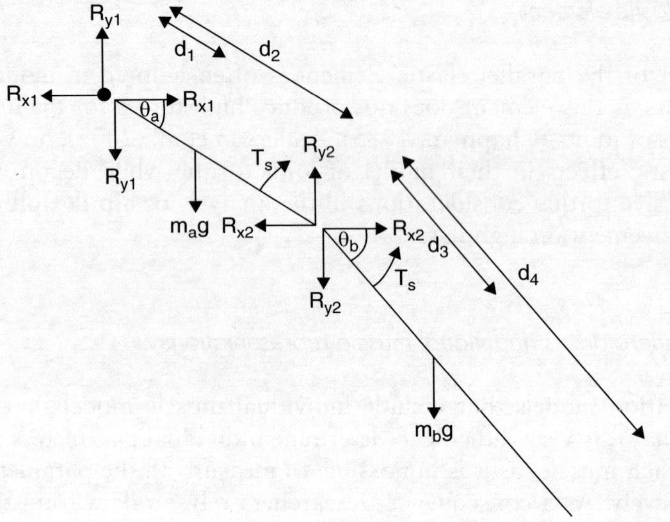

Figure 11.1 Free-body diagram of a two-segment model of a gymnast swinging around a high bar.

Generating the equations of motion

The equations of motion for a mechanical system can be generated from first principles using Newton's second law for relatively simple models with only a few segments (e.g. Hiley and Yeadon, 2003b). For a planar link model, three equations of motion are available for each segment using Newton's second law (force = mass × acceleration, $F = ma$) in two perpendicular directions and taking moments (torque = moment of inertia × angular acceleration, $T = I\alpha$) for each segment. In Figure 11.1 this allows the calculation of one angle and two reaction forces for each segment.

For more complex models a computer package is recommended, as it can take a long time to generate the equations of motion by hand and the likelihood of making errors is high. There are several commercially available software packages, for example DADS, ADAMS, AUTOLEV and SD Fast, that can generate equations of motion for a user-defined system of rigid and elastic elements. Each package allows the user to input a relatively simple description of the model and the equations of motion are then automatically generated, solved and integrated. Note that with all packages that automatically generate equations of motion, it is important to learn how to use the specific software by building simple models and performing checks to ensure that the results are correct. Some packages, such as AUTOLEV, generate computer source code, typically FORTRAN or C, for the mechanical system. The advantage of this is that the user can then customise the basic simulation model to incorporate muscle models or an optimisation routine, for example. Other more complex packages do not give full access to the source code and this can prevent the model from being customised for specific tasks.

Model input and output

Two sets of input are required for a simulation to run. First, there are the initial kinematics, which includes the mass centre velocity, and the orientation and angular velocity of each segment. The initial kinematics can be obtained from recorded performances, although it can be difficult to obtain accurate velocity estimates (Hubbard and Alaways, 1989). Second, there is information required during the simulation. A kinematically driven model requires joint angle-time histories (Yeadon, 1990a), while a kinetically driven model requires activation histories for each actuator (muscle or torque generator) in the model (Alexander, 1990; Neptune and Hull, 1999).

The output from both types of simulation model includes time histories of all the variables calculated in the simulation model. For a kinematically driven model this is the whole-body orientation, linear and angular momentum and joint torques, while for a kinetically driven model it includes the whole body orientation, linear and angular momentum and joint angle-time histories.

Integration

Running a simulation to calculate how a model moves requires a method for integrating the equations of motion over time. The simplest method to increment a set of equations of motion (ordinary differential equations) through a time interval *dt* is to use derivative information from the beginning of the interval. This is known as the 'Euler method' (Press et al., 1988):

$$x_{n+1} = x_x + \dot{x}_n dt + \tfrac{1}{2}\ddot{x}_n dt^2$$

where x, $\dot{x}$ and $\ddot{x}$ are, respectively, the displacement, velocity and acceleration and suffices n and $n + 1$ denote the nth and $(n + 1)$ step separated by a step length *dt*. The Euler method assumes a fixed step length of *dt*, where *dt* is equal to 0.0001 s, for example. The disadvantage of the Euler method is that a comparatively small step size is needed and the method is not very stable (Press et al., 1988). A better method is to use a fourth order Runge–Kutta in which four evaluations of the function are calculated per step size (Press et al., 1988). In addition, most good integration routines include a variable step size with the aim to have some predetermined accuracy in a solution with minimum computational effort (Press et al., 1988).

A kinetically driven model requires the force or torque produced by each actuator to be input to the model at each time step. The force or torque produced is a function of the actuator's activation, length and velocity. The movement of the contractile element or series elastic element must, therefore, be calculated. Caldwell (2014) gives an in-depth account of this procedure, but essentially at each time step the total length of the actuator is split between the contractile element and series elastic element in such a way that the force or torque in each element are equal.

Error checking

Whatever method is used to generate the equations of motion, it is always important that checks are carried out to ensure that no simple programming errors have been made. Examples are: energy is conserved if all damping is removed and all the muscles are switched off; the mass centre of the model follows a parabola if the forces between the simulation model and the external surface are set to zero; impulse equals change in linear momentum; angular momentum about the mass centre is conserved during flight.

Optimisation

Simulation models can be used to find the optimum technique for a specific task by running many simulations with different inputs. To perform an optimisation is a four-stage process. First, an objective function, or performance score, must be formulated which can be maximised or minimised by varying inputs to the model within realistic limits. For jumping simulations the objective function can simply be the jump height or jump distance, but for movements in which rotation is also important, a more complex function incorporating both mass centre movement and rotation is required. The challenge for formulating such an objective function is to determine appropriate weightings for each variable in the function since the weightings affect the solution. Moreover the choice of objective function can lead to very different optimum solutions (Hiley and Yeadon, 2012a).

Second, realistic limits need to be established for each of the variables, typically activation parameters to each muscle and initial conditions. Additionally the activation patterns of each muscle need to be defined using only a few parameters to keep the optimisation run time reasonably low and increase the likelihood of finding a global optimum.

Third, an algorithm capable of finding the global optimum, rather than a local optimum, is needed. Of the many algorithms available, the Simplex algorithm (Nelder and Mead, 1965), the Simulated Annealing algorithm (Goffe et al., 1994) and Genetic algorithms (van Soest and Casius, 2003) have proved popular. The Simplex algorithm typically finds a solution quickly but can get stuck at a local optimum as it only accepts downhill solutions, whereas the Simulated Annealing and Genetic algorithms are better at finding the global optimum, as they can escape from local optima.

Fourth, constraints need to be taken into account such as anatomical constraints and robustness to perturbations. Anatomical constraints, such as range of joint motion, can be adhered to by adding penalties to the objective function for simulations that exceed anatomical limits (Wilson et al., 2011) and perturbations to joint torque activation timings can be incorporated in the optimisation process to ensure that a robust optimum solution is found (Wilson et al., 2007). Constraints and robustness requirements may effectively define the movement without the need for an optimization criterion (Hiley and Yeadon, 2013).

Summary of model construction

- Decide which factors are important.
- Decide upon the number of segments and joints.
- Decide whether to include wobbling masses.
- Draw the free body diagram showing all the forces acting on the system.
- Decide whether the model is to be angle driven or torque driven.
- Decide which muscles should be represented.
- Decide how to model the interface with the ground or equipment.
- Decide whether to use a software package or to build the model from first principles.

Parameter determination

Determining parameters for a simulation model is difficult but vital, as the values chosen can have a large influence on the resulting simulations. Parameters are needed for the rigid and wobbling mass segments, muscle–tendon complexes, and viscoelastic elements in the model. Fundamentally there are two different ways to approach this, either to estimate values from the literature or take measurements on an individual to determine individual-specific parameters. There is a clear advantage to determining individual-specific parameters as it allows a model to be evaluated by comparing simulation output with performance data on the same individual.

Inertia parameters

Accurate segmental inertia values are needed for each segment in the simulation model. For a rigid segment the inertia parameters consist of the segmental mass, length, mass centre location and moment of inertia; one moment of inertia value is needed for a planar model, while three moment of inertia values are needed for a three-dimensional model. For a wobbling mass segment there are twice as many inertia parameters needed since a wobbling mass segment comprises two rigid bodies connected by viscoelastic springs.

There are two methods of obtaining rigid segmental inertia parameters. The first is to use regression equations (Hinrichs, 1985; Yeadon and Morlock, 1989) based upon anthropometric measurements and inertia parameters determined from cadaver segments (Dempster, 1955; Chandler et al., 1975). The disadvantage of this method is that the accuracy is dependent on how well the morphology of the participant compares with the cadavers used in the study. A better method, which only requires density values from cadaver studies, is to take anthropometric measurements on the participant and use a geometric model (Jensen, 1978; Hatze, 1980; Yeadon, 1990b) to determine the segmental inertia parameters. Although it is difficult to establish the accuracy of these geometric models for determining segmental inertia parameters, error values of around 2 per cent have been reported for total body mass (Yeadon, 1990b).

An alternative method that is worthy of mention is to use medical imaging techniques (Martin et al., 1989; Zatsiorsky et al., 1990a) to determine segmental inertia parameters. With current technology and ethical issues this approach is not a real alternative at present, but in the future it might provide a means for determining individual-specific segmental density values or provide a means for evaluating other methods for determining individual-specific segmental inertia parameters.

Including wobbling mass segments within the model increases the number of unknown parameters that are needed for each segment. The combined segmental inertia parameters can be calculated using a geometric model or regression equations. However, the calculation of the inertia parameters of the separate fixed and wobbling masses requires additional information on the ratio of bone to soft tissue, which is typically obtained from cadaver dissection studies (Clarys and Marfell-Jones, 1986). These ratio data can then be scaled to the specific individual using total body mass and percentage body fat (Pain and Challis, 2006; Wilson et al., 2006). In the future it may be possible to improve this method by determining the inertia parameters for the rigid and wobbling masses of each segment directly from medical imaging.

Strength parameters

Determining accurate individual-specific strength parameters for muscle–tendon complexes is a major challenge in sports biomechanics, which has resulted in two different ways to represent the forces produced by muscles. The first is to include all the major muscles that cross a joint in the simulation model as individual muscle–tendon complexes, with the parameters for the individual muscles obtained mainly from animal experiments (e.g. Gerritsen et al., 1995). Although the parameters are sometimes scaled to an individual or group of individuals based upon either isometric measurements (Hatze, 1981b) or isovelocity measurements (Conceição et al., 2012), this method does not give a complete set of individual-specific strength parameters. The alternative approach is to use torque generators at each joint in the model to represent the effect of all the muscles around a joint (flexors and extensors represented by separate torque generators). The advantage of this approach is that the net torque at a given joint can be measured on a constant velocity dynamometer over a range of joint angular velocities and joint angles for the participant and so individual-specific parameters can be determined that define maximal voluntary torque as a function of muscle angle and velocity (King and Yeadon, 2002; Yeadon et al., 2006b; Forrester et al., 2011). If the athlete is not available, individual-specific strength parameters for each torque generator can be estimated by matching recorded performances using a simulation model (King et al., 2009). Torque-driven models still require data from the literature to determine the parameters for the series elastic element for each torque generator. In recent studies (King et al., 2006) it has been assumed that the series elastic element stretches by 5 per cent of its resting length during isometric contractions (De Zee and Voigt, 2001; Muramatsu et al., 2001). Although it would be desirable to be able to determine series elastic element parameters directly from measurements on an individual, it has previously been shown that simulation results were not sensitive to these parameter values (Yeadon and King, 2002).

Viscoelastic parameters

Viscoelastic parameters are required for springs that are included within a simulation model: connection of wobbling masses, shoulder joint, foot (or hand) ground interface and equipment. Sometimes these springs represent specific elements where it is possible to determine viscoelastic properties from measurements (Pain and Challis, 2001b) while in other models the springs represent more than one viscoelastic element and so make it much harder to determine the parameters from experiments (e.g. Yeadon and King, 2002). Viscoelastic parameters should ideally be determined from independent tests and then fixed within the model for all simulations (Gerritsen et al., 1995; Pain and Challis, 2001a; Glynn et al., 2011). If this is not possible, the viscoelastic parameters can be determined through an optimisation procedure by choosing initial values and then allowing the parameters to vary within realistic bounds until an optimum match between simulation and performance is found (King et al., 2011). With this method a torque-driven or angle-driven simulation model can be used, although it is easier to implement in an angle-driven model as the joint angle changes are specified and so there are fewer parameters to be determined. Optimising the parameter values has the potential for the springs to compensate for errors in the model. This can be overcome by using a small set of spring parameters, determining the parameters from more than one trial and then fixing the parameter values for the model evaluation. For example, Yeadon and King (2002) determined viscoelastic parameters for the interface between foot and tumble track from one trial using a torque-driven model and then evaluated the model on a different trial, while Yeadon et al. (2006a) determined viscoelastic parameters for the interface of a diver with a springboard from four trials using an angle-driven model. Using more than one trial for determining the spring parameters also has the advantage that the model output should not be overly sensitive to the parameter values used.

Model evaluation

Model evaluation is an essential step in the process of developing a simulation model and should be carried out before a model is used in applications. Although this step was identified as an important part of the process over 25 years ago (Panjabi, 1979) the weakness of many simulation models is still that their accuracy is unknown (Yeadon and Challis, 1994). While a number of models have been evaluated to some extent, such as those of Hatze (1981a), Yeadon et al. (1990a), Neptune and Hull (1998), Brewin et al. (2000), Fujii and Hubbard (2002), Yeadon and King (2002), Hiley and Yeadon (2003a),

King et al. (2006), Sheets and Hubbard (2008), Allen et al. (2010), Yeadon et al. (2010), Kentel et al. (2011) and King et al. (2011), many have not been evaluated at all.

The complexity of the model and its intended use should be taken into account when evaluating a model. For a simple model (e.g. Alexander, 1990), which is used to make general predictions, it may be sufficient to show that results are of the correct magnitude. In contrast, if a model is being used to investigate the factors that determine optimum performance in jumping, the model should be evaluated quantitatively so that the accuracy of the model is known (e.g. King et al., 2006). Ideally the model evaluation should encompass the range of initial conditions and activities that the model is used for, with little extrapolation of the model to cases in which the accuracy is unknown (Panjabi, 1979). For example, if a simulation model of springboard diving is evaluated successfully for forward dives, the model may not work for reverse dives and so it should be also evaluated using reverse dives.

The purpose of model evaluation is to determine the accuracy, which can then be borne in mind when considering the results of simulations. Furthermore a successful evaluation gives confidence that the model assumptions are not erroneous and that there are no gross modelling defects or simulation software errors. Ideally the evaluation process should include all aspects of the model that are going to be used to make predictions. If a model is going to be used to investigate the effect of initial conditions on maximum jump height then the model should be evaluated quantitatively to show that, for a given set of initial conditions, the model can perform the movement in a similar way and produce a similar jump height. If a model is to be used to examine how the knee flexor and extensor muscles are used in jumping, the model should be evaluated to show that for a given jump the model uses similar muscle forces to the recorded performance.

To evaluate a simulation model is challenging and may require several iterations of model development before the model is evaluated satisfactorily. Initially data must be collected on a performance by the sports participant. Ideally this should be an elite performer who is able to work maximally throughout the testing and produce a performance that is close to optimal. Time histories of kinematic variables, from video or an automatic system, kinetic variables from force plate or force transducers, and electromyogram (EMG) histories, if possible, should be obtained. Individual-specific model parameter values, such as anthropometry and strength, are then determined from the measurements taken on the individual, with as little reliance on data from the literature as possible (Wilson et al., 2006; Yeadon et al., 2006b). The initial kinematic conditions (positions and velocities) for the model are then determined from the performance data and input to the model along with any other time histories that are required for the model to run a single simulation. If the model is kinetically driven this will consist of the activation-time history for each actuator (Yeadon and King, 2002), while if the model is kinematically driven the time history of each joint angle will be required (Hiley and Yeadon, 2003a). Once a single simulation has been run, a difference score should be calculated by quantitatively comparing the simulation with the recorded performance. The formulation of the score depends on the activity being simulated, but it should include all features of the performance that the model should match, such as joint angle changes, linear and angular momentum, and floor movement. The difficulty in combining severable variables into one score is that appropriate weightings need to be chosen for each part of the objective function. For example, Yeadon and King (2002) assumed that a 1° difference in a joint angle at take-off was equivalent to a 1 per cent difference in mass centre velocity at take-off. Furthermore, for variables that cannot be measured accurately, such as wobbling mass movement, it may be more appropriate to add a penalty to the difference score if too much movement occurs (King et al., 2006). Finally the input to the model is then varied until the best comparison is found (score minimised) using an optimisation routine. If the comparison between performance and simulation is close (Figure 11.2) then the model can be used to run simulations. If not, then the model complexity or model parameters need to be modified and the model re-evaluated. If the comparison gives a percentage difference of less than 10 per cent this is often sufficient for applications in sports biomechanics.

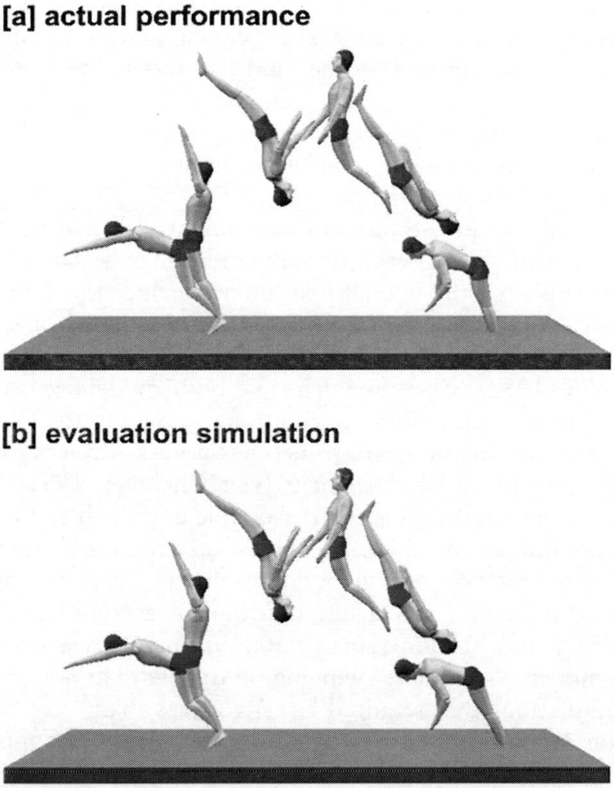

Figure 11.2 Comparison of performance and simulation graphics for the tumbling model of Yeadon and King (2002).

Issues in model design

The design of a particular model should be driven by the intended use and the questions to be answered. For example, if the aim is to determine the forces that act within the human body during running, then an inverse dynamics model may be more appropriate than a forward dynamics model. If the aim is to demonstrate some general mechanical principles for a type of movement then a simple model may be adequate. The issue of model complexity is not simple, however. While it is evident that simple models such as Alexander's (1990) model of jumping can give insight into the mechanics of technique, there is often a tendency to rely on the quantitative results without recourse to model evaluation. The issue of model evaluation for a simple model is problematic since all that can be realistically expected is a ballpark or order-of-magnitude accuracy. Achieving anything approaching 10 per cent accuracy when compared with performance usually requires a model of some complexity, comprising several segments, realistic joint drivers and elastic elements. The development of such a model is a non-trivial endeavour. Sprigings and Miller (2004, p. 287) argue the case for 'the use of the simplest possible model capable of capturing the essence of the task being studied', citing Alexander (1990) and Hubbard (1993) in support. The problem here is deciding at what point a model is too simple. If a model is so simple that it is 30 per cent inaccurate then it is difficult to justify conclusions indicated by the model results unless they are robust to a 30 per cent inaccuracy. It is evident that some measure of model accuracy is needed to reach conclusions.

Simple models of throwing in which the implement is modelled as an aerodynamic rigid body (Hubbard and Alaways, 1987) need to be complemented by a representation of the ability of the thrower to impart velocity in a given direction (Hubbard et al., 2001) so that realistic simulations may be carried out. The same considerations apply to other models that do not include the human participant.

Although a rigid body may be adequate for a model of equipment, it is likely to be too simple for a model of an activity such as high jumping (Hubbard and Trinkle, 1985a, b), although a rigid body model has been used to give insight into the two general modes of rotational aerial motion (Yeadon, 1993a).

Joint angle time histories are sometimes used as drivers for a simulation model. In the case of aerial movement (van Gheluwe, 1981; Yeadon et al., 1990a) it can be argued that this is a reasonable approach so long as the angular velocities are limited to achievable values. In activities where there are large contact forces with the external surroundings this approach is more problematic since steps need to be taken to ensure that the corresponding joint torques are achievable. Hiley and Yeadon (2003a; 2003b) and Brewin et al. (2000) used angle-driven models to simulate swinging on the high bar and on the rings and eliminated simulations that required larger torques than were achieved by the participant on a constant velocity dynamometer. Another approach is to use joint torques as drivers, where the maximum voluntary joint torque is a function of angular velocity (Alexander, 1990) and, possibly, of joint angle (King and Yeadon, 2003). This approach leads to more realistic simulations than the use of angle-driven models but there is a corresponding loss of the simple control of joint angles. Finally there are models that use representations of individual muscles or muscle groups crossing a joint (Hatze, 1981a; Neptune and Hull, 1998) and these have the potential to provide even more accurate representations but pose the problem of determining appropriate muscle parameter values. A torque-driven model may be preferred to a muscle-driven model as it allows individual-specific strength parameters to be determined. When estimates of internal loading are required, however, a muscle-driven model may be more appropriate (Mills et al., 2008; Mills et al., 2009).

Reviews of computer simulation modelling are provided by Miller (1975), King (1984), Vaughan (1984), Yeadon (1987), Vaughan (1989), Hubbard (1993), Alexander (2003) and King and Yeadon (2015).

THE INVERSE DYNAMICS PROBLEM

The inverse dynamics problem is to determine the forces that must act to produce a given motion. Theoretically the only information needed includes the time histories of the variables that define the motion of the system. From a practical perspective, however, estimates of angular accelerations from the given data typically have large errors and so additional information is often provided in the form of recorded ground reaction forces. As an example, a four-segment model representing a handstand on a force plate will be used to determine the torques acting at each joint.

An inverse dynamics model of a handstand

The body is represented by four rigid segments H, A, B, C representing the hands, arms, trunk+head and legs (Figures 11.3 and 11.4). Newton–Euler equations are used to generate three equations per segment for the six joint reaction forces, three angular accelerations and three joint torques. This system of 12 equations in 12 unknowns is reduced to a system of six equations in the joint accelerations and joint torques by eliminating the six reaction forces. A knowledge of the segmental inertia parameters of a gymnast together with the time histories of the three joint angles during a handstand then permits the calculation of the joint torque time histories.

Each of the four segments in Figure 11.3 – Hand (H), Figure 11.3(a); Arm (A), Figure 11.3(b); Body (B), Figure 11.3(c); Leg (C), Figure 11.3(d) – produces three equations: one for resultant vertical force, one for resultant horizontal force, one for moments about the mass centre.

For Figure 11.3(a):

$$\uparrow: R - R_1 - m_h g = 0 \tag{11.1}$$

$$\rightarrow: F - F_1 = 0 \tag{11.2}$$

$$H: -T_1 + R(x_p - x_h) + R_1(x_h - x_1) + F(z_h - z_p) + F_1(z_1 - z_h) = 0 \tag{11.3}$$

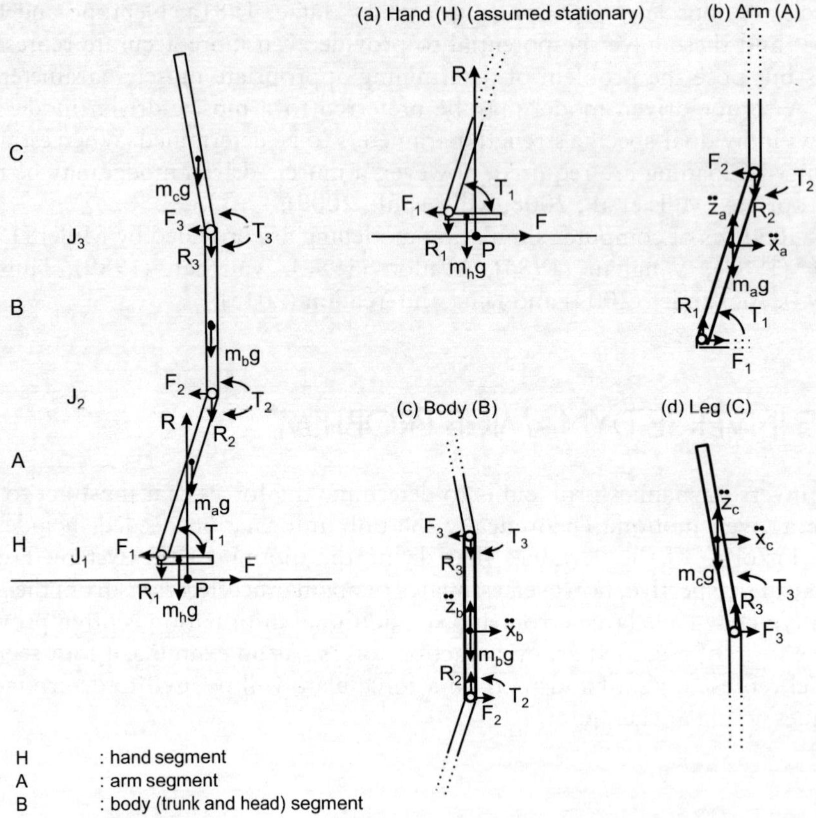

H	: hand segment
A	: arm segment
B	: body (trunk and head) segment
C	: leg segment
J_1	: wrist joint
J_2	: shoulder joint
J_3	: hip joint
(x_i, z_i)	: joint centre coordinates (i = 1, 3)
F_i	: horizontal joint reaction forces (i = 1, 3)
R_i	: vertical joint reaction forces (i = 1, 3)
T_j	: torque about joint centre J_i (i = 1, 3)
P	: centre of pressure
F	: horizontal reaction force on H
R	: vertical reaction force on H
(x_j, z_j)	: segment mass centre coordinates (j = h, a, b or c)
(x_p, z_p)	: point of force application (z_p is assumed = 0)
$\ddot{x}_j$	: horizontal linear accelerations of segment mass centre (j = h, a, b or c)
$\ddot{z}_j$	: vertical linear accelerations of segment mass centres (j = h, a, b or c)
m_j	: segment mass (j = h, a, b or c)
I_j	: moment of inertia about segment mass centres (j = h, a, b or c)
ϕ_j	: segment angular accelerations (j = h, a, b or c)
g	: acceleration due to gravity

Figure 11.3 Free-body diagram for a four-segment model of a handstand.

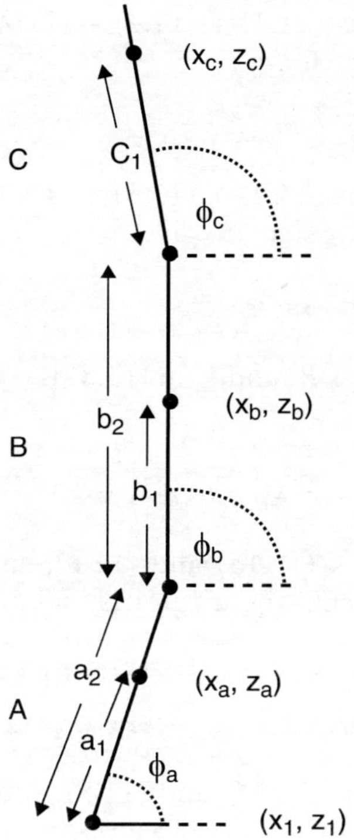

Figure 11.4 Four-segment model of a handstand.

For Figure 11.3(b):

$$\uparrow: R_1 - R_2 - m_a g = m_a \ddot{z}_a \quad (11.4)$$

$$\rightarrow: F_1 - F_2 = m_a \ddot{x}_a \quad (11.5)$$

$$A: -T_2 + T_1 + F_1(z_a - z_1) + F_2(z_2 - z_a) - R_1(x_a - x_1) - R_2(x_2 - x_a) = I_a \ddot{\varphi}_a \quad (11.6)$$

For Figure 11.3(c):

$$\uparrow: R_2 - R_3 - m_b g = m_b \ddot{z}_b \quad (11.7)$$

$$\rightarrow: F_2 - F_3 = m_b \ddot{x}_b \quad (11.8)$$

$$B: T_2 - T_3 + F_2(z_b - z_2) + F_3(z_3 - z_b) - R_2(x_b - x_2) - R_3(x_3 - x_b) = I_b \ddot{\varphi}_b \quad (11.9)$$

For Figure 11.3(d):

$$\uparrow: R_3 - m_c g = m_c \ddot{z}_c \quad (11.10)$$

$$\rightarrow: F_3 = m_c \ddot{x}_c \quad (11.11)$$

$$C: T_3 + F_3(z_c - z_3) - R_3(x_c - x_3) = I_c \ddot{\varphi}_c \quad (11.12)$$

Combining equations (11.1) + (11.4) + (11.7) + (11.10) resolves forces vertically for the whole system, to give:

$$R - mg = m_a \ddot{z}_a + m_b \ddot{z}_b + m_c \ddot{z}_c \tag{11.13}$$

Combining equations (11.2) + (11.5) + (11.8) + (11.11) resolves forces horizontally for the whole system, to give:

$$F = m_a \ddot{X}_a + m_b + \ddot{X}_b + m_c \ddot{X}_c \tag{11.14}$$

Substituting for values R_1 and F_1 in (11.3) is equivalent to taking moments about J_1 for H and gives:

$$T_1 = F(Z_1 - Z_p) + R(x_p - x_1) - m_h g(x_h - x_1) \tag{11.15}$$

Combining equations (11.15) and (11.6), substituting for R_2 and F_2 is equivalent to taking moments about J_2 for H and A and gives:

$$\begin{aligned} F(z_2 - z_p) - R(x_2 - x_p) + m_h g(x_2 - x_h) + m_a g(x_2 - x_a) \\ = T_2 + I_a \ddot{\varphi}_a + m_a \ddot{x}_a (z_2 - z_a) - m_a \ddot{z}_a (x_2 - x_a) \end{aligned} \tag{11.16}$$

Combining equations (11.16) and (11.9), substituting for R_3 and F_3 and taking moments about J_3 for H, A and B gives:

$$\begin{aligned} F(z_3 - z_p) - R(x_3 - x_p) + m_h g(x_3 - x_h) + m_a g(x_3 - x_a) + m_b g(x_3 - x_b) \\ = T_3 + I_a \ddot{\varphi}_a + I_b \ddot{\varphi}_b + m_a \ddot{x}_a (z_3 - z_a) + m_a \ddot{x}_b (z_3 - z_b) \\ - m_a \ddot{z}_a (x_3 - x_a) - m_b \ddot{z}_b (x_3 - x_b) \end{aligned} \tag{11.17}$$

Combining equations (11.17) and (11.12) is equivalent to taking moments about P for the whole system and gives:

$$\begin{aligned} -mg(x - x_p) = I_a \ddot{\varphi}_a + I_b \ddot{\varphi}_b + I_c \ddot{\varphi}_c - m_a \ddot{x}_a (z_a - z_p) - m_b \ddot{x}_b (z_b - z_p) \\ - m_c \ddot{x}_c (z_c - z_p) + m_a \ddot{z}_a (x_a - x_p) + m_b \ddot{z}_b (x_b - x_p) \\ + m_c \ddot{z}_c (x_c - x_p) \end{aligned} \tag{11.18}$$

Therefore, eliminating reactions at joints has left six equations of motion. Using the representation of Figure 11.4, the geometric equivalents of segment mass centre linear accelerations can be obtained by differentiating the position values twice.

$$\begin{aligned}
x_a &= x_1 + a_1 \cos \varphi_a & \ddot{x}_a &= -a_1 \sin \varphi_a \ddot{\varphi}_a - a_1 \cos \varphi_a \dot{\varphi}_a^2 \\
z_a &= z_1 + a_1 \sin \varphi_a & \ddot{z}_a &= -a_1 \cos \varphi_a \ddot{\varphi}_a - a_1 \sin \varphi_a \dot{\varphi}_a^2 \\
x_b &= x_1 + a_2 \cos \varphi_a & \ddot{x}_b &= -a_2 \sin \varphi_a \ddot{\varphi}_a - \mathrm{a}_2 \cos \varphi_a \dot{\varphi}_a^2 \\
&\quad + \mathrm{b}_1 \cos \varphi_b & &\quad - \mathrm{b}_1 \sin \varphi_b \ddot{\varphi}_b - \mathrm{b}_1 \cos \varphi_b \dot{\varphi}_b^2 \\
z_b &= z_1 + a_2 \sin \varphi_a & \ddot{z}_b &= a_2 \cos \varphi_a \ddot{\varphi}_a - \mathrm{a}_2 \sin \varphi_a \dot{\varphi}_a^2 \\
&\quad + \mathrm{b}_1 \sin \varphi_b & &\quad + \mathrm{b}_1 \cos \varphi_b \ddot{\varphi}_b - \mathrm{b}_1 \sin \varphi_b \dot{\varphi}_b^2
\end{aligned}$$

$$x_c = x_1 + a_2 \cos\varphi_a + b_2 \cos\varphi_b + c_1 \cos\varphi_c \qquad \ddot{x}_c = -a_2 \sin\varphi_a \ddot{\varphi}_a - a_2 \cos\varphi_a \dot{\varphi}_a^2 - b_2 \sin\varphi_b \ddot{\varphi}_b - b_2 \cos\varphi_b \dot{\varphi}_b^2 - c_1 \sin\varphi_c \ddot{\varphi}_c - c_1 \cos\varphi_c \dot{\varphi}_c^2$$

$$z_c = z_1 + a_2 \sin\varphi_a + b_2 \sin\varphi_b + c_1 \sin\varphi_c \qquad \ddot{z}_c = -a_2 \cos\varphi_a \ddot{\varphi}_a - a_2 \sin\varphi_a \dot{\varphi}_a^2 - b_2 \cos\varphi_b \ddot{\varphi}_b - b_2 \sin\varphi_b \dot{\varphi}_b^2 - c_1 \cos\varphi_c \ddot{\varphi}_c - c_1 \sin\varphi_c \dot{\varphi}_c^2$$

By substituting the geometric equivalents in place of the linear acceleration terms in equations (11.13) through (11.18) and re-arranging terms, we obtain six linear equations in the following form to solve for six unknowns (T_1, T_2, T_3, $\phi_{a,}$ $\phi_{b,}$ ϕ_c).

$$\begin{aligned}
A_{11}T_1 + A_{12}T_2 + A_{13}T_3 + A_{14}\ddot{\varphi}_a + A_{15}\ddot{\varphi}_b + A_{16}\ddot{\varphi}_c &= B_1 \\
A_{21}T_1 + A_{22}T_2 + A_{23}T_3 + A_{24}\ddot{\varphi}_a + A_{25}\ddot{\varphi}_b + A_{26}\ddot{\varphi}_c &= B_2 \\
A_{31}T_1 + A_{32}T_2 + A_{33}T_3 + A_{34}\ddot{\varphi}_a + A_{35}\ddot{\varphi}_b + A_{36}\ddot{\varphi}_c &= B_3 \\
A_{41}T_1 + A_{42}T_2 + A_{43}T_3 + A_{44}\ddot{\varphi}_a + A_{45}\ddot{\varphi}_b + A_{46}\ddot{\varphi}_c &= B_4 \\
A_{51}T_1 + A_{52}T_2 + A_{53}T_3 + A_{54}\ddot{\varphi}_a + A_{55}\ddot{\varphi}_b + A_{56}\ddot{\varphi}_c &= B_5 \\
A_{61}T_1 + A_{62}T_2 + A_{63}T_3 + A_{64}\ddot{\varphi}_a + A_{65}\ddot{\varphi}_b + A_{66}\ddot{\varphi}_c &= B_6
\end{aligned}$$

All of the terms held in the coefficients A_{11} through B_6 can be derived from video or force data at each instant in time. A linear equation solver is used to determine estimates for the six unknowns at each time instant. However, some of the equation coefficients involve cosϕ_a, cosϕ_b, cosϕ_c which result in singularities in the calculated torques and angular accelerations around $\phi_j = 90°$ (j = a, b, c). To avoid this problem a further three equations are added using video estimates e_1, e_2, e_3 of the angular accelerations $\ddot{\varphi}_a$, $\ddot{\varphi}_b$, $\ddot{\varphi}_c$. These may be written as:

$$\begin{aligned}
A44\ddot{\varphi}_a &= A44e1 \\
A55\ddot{\varphi}_b &= A55e2 \\
A66\ddot{\varphi}_c &= A66e3
\end{aligned}$$

which match the coefficients of $\ddot{\varphi}_a$, $\ddot{\varphi}_b$, $\ddot{\varphi}_c$ in the last three of the six previous equations. This gives an over-determined system of nine equations for the six unknowns and a least-squares equation solver results in solutions without singularities. The addition of the further three linear equations constrains the angular acceleration estimates returned by the solver to sensible values and consequently the torque values returned are also more stable (Figure 11.5).

In analysing movements with an impact phase, inverse dynamics is more problematic since it is not possible to include wobbling masses within an inverse dynamics model. In such movements, a constrained forward dynamics model is a better way to proceed.

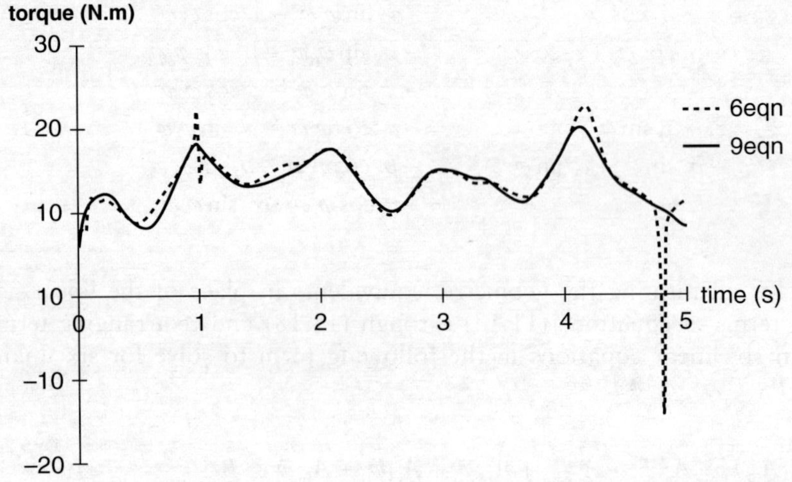

Figure 11.5 Joint torque obtained by inverse dynamics using six-equation system and nine-equation over-determined system (Reproduced from Yeadon, M.R. and Trewartha, G. 2003. 'Control strategy for a hand balance'. *Motor Control, 7*, p. 418 by kind permission of Human Kinetics).

Solving the inverse dynamics problem using forward dynamics simulation

An alternative to inverse dynamics is to use a constrained (angle-driven) forward dynamics simulation model for solving the inverse dynamics problem. This method allows wobbling masses to be included within the model, which can have a substantial effect on the joint torques calculated, especially during impacts (Pain and Challis, 2006). The disadvantage of using a forward dynamics formulation is that it may be necessary to optimise model parameters to find a solution that matches the recorded performance. In addition it is not possible to take advantage of an over-determined system and accurate acceleration values are needed, which can be almost impossible to calculate during impacts.

With a constrained forward dynamics planar model there are three degrees of freedom for whole body motion – horizontal and vertical translation of the mass centre and whole body orientation – along with three degrees of freedom for each wobbling mass segment in the model. Time histories of the joint angles and external forces are input to the model and the motion of the model is calculated along with the joint torques required to satisfy the joint angle changes. King et al. (2003) used this method to calculate the net joint torque at the knee for the take-off phase in a jump for height. The peak knee torques calculated using quasi-static, pseudo-inverse dynamics (no wobbling mass movement), and constrained forward dynamics were 747 N m, 682 N m and 620 N m. Including segment accelerations resulted in lower peak values and the inclusion of wobbling masses resulted in a smoother knee torque time history. However, all the calculated peak knee torques were in excess of the eccentric maximum that could be exerted by the participant, which was estimated to be 277 N m from constant velocity experiments with the participant (Figure 11.6). King et al. (2003) concluded that the discrepancy requires further investigation but it is

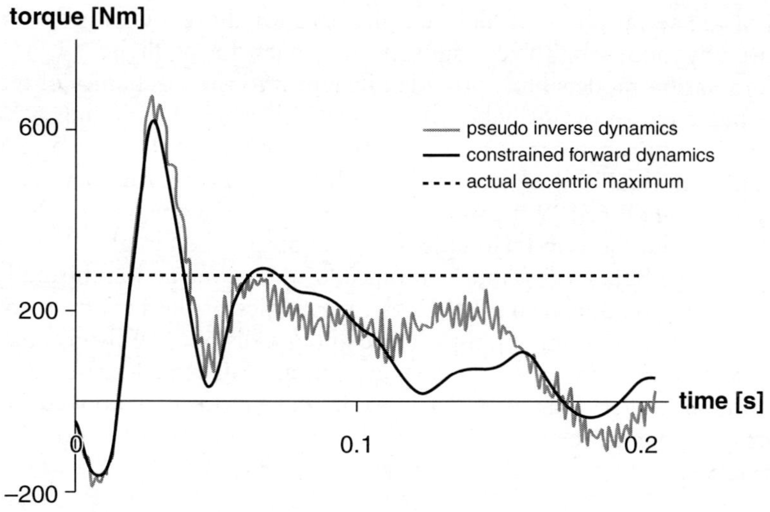

Figure 11.6 Knee joint torque calculated using pseudo inverse dynamics and constrained forward dynamics.

likely to be due to modelling the knee as a simple frictionless pin joint or may be a consequence of errors in the digitised data. Indeed it has been shown with an angle-driven simulation model of triple jump take-offs that the accurate calculation of internal forces requires a model that incorporates compliance elsewhere in the link system (Allen et al., 2012).

APPLICATIONS

In this section, various examples of modelling in sports biomechanics are given to illustrate model implementation and address the problems of optimisation and the control of sports movements.

Understanding the mechanics of sports technique

It is possible to use a simulation model to gain insight into the mechanics of sports techniques. Kinematic and kinetic data on a performance may suggest that a particular technique is responsible for the outcome but without some method of quantifying contributions little can be concluded. With a simulation model the efficacy of various techniques may be evaluated and so give insight into what really produces the resulting motion. Van Gheluwe (1981) and Yeadon (1993b; 1993c) used angle-driven simulation models of aerial movement to investigate the capabilities of various contact and aerial twisting techniques. They found that twist could be produced in the aerial phase of a plain somersault using asymmetrical movements of arms or hips. Dapena (1981) used an angle-driven model of the aerial phase of high jumping to show how a greater height could be cleared by modifying the configuration changes during flight.

Simulation models have provided insight into the mechanics of technique in: the flight phase of springboard diving (Miller, 1971), circling a high bar (Arampatz and Brüggemann, 1999; Arampatzis et al., 2001; Hiley and Yeadon, 2001; Hiley et al., 2008), skateboarding (Hubbard, 1980), the curved approach in high jumping (Tan, 1997), landings in gymnastics (Requejo et al., 2004; Mills et al., 2010), double arm technique in triple jump (Allen et al., 2010), circling the asymmetric bars (Hiley and Yeadon, 2007), co-contraction during landings from a height (Yeadon et al., 2010), the mechanics of the Fouette turn (Imura and Yeadon, 2010), running jumps for height and distance (Wilson et al., 2011), and club position relative to the swing plane in golf (MacKenzie, 2012). It is important, however, that a model is evaluated before it is applied, since the insights gained may be into the (possibly incorrect) model rather than into performance.

Contributions

Simulation models may be used to determine the contributions of various aspects to the overall performance by simulating the effect of what happens when an aspect is removed or when just one aspect is present. So that a variable can be used to quantify 'contributions', it is necessary that such measures are additive. For example, in twisting somersaults the use of the final twist angle achieved can lead to problems since the sum of twist angles produced by various techniques is likely to be greater than the twist resulting from the concurrent use of all the techniques. Additionally, technique in the latter part of a twisting somersault may be primarily directed towards stopping the twist rather than producing the twist. Because of these effects, Yeadon (1993d) used the maximum tilt angle as a measure of the twist potential in a movement. The tilt angles calculated in this way were additive and could be sensibly referred to as contributions from various twisting techniques. This method has been applied to twisting technique in gymnastics dismounts (Yeadon et al., 1990b; Yeadon, 1994; 1997), springboard diving (Yeadon, 1993e) and freestyle skiing (Yeadon, 1989).

Brewin et al. (2000) used a model of a gymnast and the rings apparatus to determine the contributions of technique and the elasticity of the gymnast and rings apparatus to the reduction of loading at the shoulders. It was found that technique reduced the loading by 2.7 body weights while elasticity reduced the loading by 1.1 body weights resulting in the actual loading of 8.5 body weights. King and Yeadon (2005) used a five-segment model of a gymnast during vaulting take-off to investigate factors affecting performance of the Hecht vault. It was found that shoulder torque made only a small contribution of 7° to the resulting rotation whereas shoulder elasticity contributed 50° to the rotation in flight.

King et al. (2012b) used a model of single-handed backhand strokes in tennis to investigate the contributions to wrist extension torque. It was found that the most important factor was ball impact location, with an off-centre impact location below the longitudinal axis of the racket resulting in six times more wrist extension torque when compared to the equivalent centre impact simulation.

Optimisation of sports technique

Since a single simulation of a sports movement can take less than a second, it may be possible to run hundreds of thousands of simulations in a single day. This opens the way to investigating optimised performance by means of a theoretical study. The technique used in a sports movement is characterised using various parameters and then an optimisation procedure is used to find the best set of parameter values that maximises or minimises some performance score. At its simplest, this could involve determining the optimal initial conditions in a projectile event such as basketball (Schwark et al., 2004) or javelin (Best et al., 1995; Hubbard and Alaways, 1987). Similarly, optimum bat swing trajectories can be determined for maximum baseball range (Sawicki et al., 2003). In such optimisations, relatively few parameters corresponding to one instant in time are optimised. In such cases it is important to take account of the interdependence of release parameters arising from the characteristics of the human participant (Hubbard et al., 2001).

More challenging are dynamic optimisations in which the time history of a sports technique is optimised. Typically this requires many parameters to characterise the technique used. In the case of angle-driven models, it is relatively simple to ensure that anatomical constraints at the joints are not violated (Hiley and Yeadon, 2003a; Yeadon, 2013). For models driven by joint torques or muscle representations, such constraints cannot be imposed directly but can be accommodated using penalties as part of the optimisation function (Kong et al., 2005). Optimisation of sports technique may involve several hundred thousand simulations taking many hours or days to run. The run time may be reduced using more computing hardware via parallel processing or by reformulating the problem using Direct Collocation (van den Bogert and Ackermann, 2009).

The performance score of the sports skill could simply be the distance thrown (Hubbard, 1984) or the distance jumped (Hatze, 1983; Hubbard et al., 1989), the height jumped (Nagano and Gerritsen, 2001; Cheng and Hubbard, 2004), the amount of rotation produced (Sprigings and Miller, 2004; Hiley and Yeadon, 2005), power output (van den Bogert, 1994), fatigue (Neptune and Hull, 1999), or more complex combinations of performance variables (Gervais, 1994; Koh et al., 2003). While such an approach may work, it is also possible that the optimum solution is sensitive to small variations in technique, leading to inconsistent performance. This issue of robustness to perturbations will be discussed in the next section.

CONTROL OF SPORTS MOVEMENTS

If a technique produces a perfect performance then inevitably there will be deviations from this performance resulting from small errors in timing. If the magnitude of such timing errors is known then the performance error can be calculated or, conversely, the timing errors may be estimated from the performance error. Yeadon and Brewin (2003) estimated that timing errors were of the order of 15 ms for a longswing to a still handstand on rings. Some movements, such as a hand balance on floor (Yeadon and Trewartha, 2003) or a non-twisting straight somersault (Yeadon and Mikulcik, 1996), may be inherently unstable and may require continual proprioceptive feedback control to be performed at all. Other movements, such as twisting somersaults (Yeadon and Hiley, 2014), may require continual feedback correction to prevent drift away from the targeted performance. Variation in approach characteristics in tumbling may be compensated for by modifications in take-off technique using feed-forward control, but only if such variation can be estimated in advance with sufficient accuracy (King and Yeadon, 2003).

Variation in technique can also be coped with by adopting a technique that is relatively insensitive, or robust, to perturbations (van Soest et al., 1994; King and Yeadon, 2004). In cases where the limits of timing a movement are close to being reached, such considerations may be the main driver for selecting technique (Hiley and Yeadon, 2003b; 2012b; 2013). In order to introduce realistic perturbations it is necessary to establish the spatial and temporal variation inherent in such movements (Hiley et al., 2013).

Feedback adjustments are dependent upon the information provided by the sensory system. In twisting somersaults it is of advantage to view the landing area during flight by adopting an appropriate head orientation (Yeadon and Knight, 2012).

CONDUCTING A STUDY

The main steps in conducting a study using a simulation model are as follows:

- identification of the research questions to be addressed
- design of the model with these aims in mind
- model construction
- data collection for model input and parameter determination
- parameter determination
- model evaluation
- experimental design of simulations to be run
- results of simulations
- conclusions: answering the research questions

REPORTING A STUDY

The format for reporting on a study will depend to some extent on the intended readership but should reflect the main steps listed in the previous section. Figures should be used when presenting a description of the model, performance data, simulation output, and model evaluation comparisons. The structure of a report or paper is usually along the following traditional lines:

- introduction: background, statement of aims
- methods: model design, parameter determination, data collection, evaluation
- results: simulation output, graphs, graphics, tables
- discussion: addressing the aims, limitations, conclusions.

SUMMARY

The use of simulation models in sport can give insight into what is happening or, in the case of a failing model, what is not happening (Niklas, 1992). Models also provide a means for testing hypotheses generated from observations or measurements of performance. It should be remembered, however, that all models are simplifications and will not reflect all aspects of the real system. The strength of computer simulation modelling in providing sports science support for performers is that it can provide general research results for the understanding of elite performance. While there is also the possibility of providing individual advice using personalised models, most sports biomechanics practitioners are a long way from realising this at present.

REFERENCES

Alexander, R. M. (1990) 'Optimum take-off techniques for high and long jumps', *Philosophical Transactions of the Royal Society of London – Series B*, 329: 3–10.

Alexander, R. M. (2003) 'Modelling approaches in biomechanics', *Philosophical Transactions of the Royal Society of London – Series B: Biological Sciences*, 358: 1429–1435.

Allen, S. J., King, M. A. and Yeadon, M. R. (2010) 'Is a single or double arm technique more advantageous in triple jumping?', *Journal of Biomechanics*, 43: 3156–3161.

Allen, S. J., King, M. A. and Yeadon, M. R. (2012) 'Models incorporating pin joints are suitable for simulating performance but unsuitable for simulating internal loading', *Journal of Biomechanics*, 45: 1430–1436.

Anderson, D. E., Madigan, M. L. and Nussbaum, M. A. (2007) 'Maximum voluntary joint torque as a function of joint angle and angular velocity: model development and application to the lower limb', *Journal of Biomechanics*, 40: 3105–3113.

Arampatzis, A. and Brüggemann, G. P. (1999) 'Mechanical energetic processes during the giant swing exercise before dismounts and flight elements on the high bar and the uneven parallel bars', *Journal of Biomechanics*, 32: 811–820.

Arampatzis, A., Brüggemann, G. P. and Klapsing, G. M. (2001) 'Mechanical energetic processes during the giant swing before the Tkatchev exercise', *Journal of Biomechanics*, 34: 505–512.

Begon, M., Yeadon, M. R. and Hiley, M. J. (2009) 'Effect of hip flexibility on optimal stalder performances on high bar', *Computer Methods in Biomechanics and Biomedical Engineering*, 12: 575–583.

Best, R. J., Bartlett, R. M. and Sawyer, R. A. (1995) 'Optimal javelin release', *Journal of Applied Biomechanics*, 11: 371–394.

Bobbert, M. F., Houdijk, J. H. P., de Koning, J. J. and de Groot, G. (2002) 'From a one-legged vertical jump to the speed-skating push-off: a simulation study', *Journal of Applied Biomechanics*, 18: 28–45.

Brewin, M. A., Yeadon, M. R. and Kerwin, D. G. (2000) 'Minimising peak forces at the shoulders during backward longswings on rings', *Human Movement Science*, 19: 717–736.

Caldwell, G. E. (2014) 'Muscle modeling', in G. E. Robertson, G. E. Caldwell, J. Hamill, G. Kamen, and S. N. Whittlesey (eds), *Research Methods in Biomechanics* (2nd edn.), pp. 203–232, Champaign, IL: Human Kinetics.

Challis, J. H. and Pain, M. T. G. (2008) 'Soft tissue motion influences skeletal loads during impacts', *Exercise and Sport Sciences Reviews*, 36: 71–75.

Chandler, R. F., Clauser, C. E., McConville, J. T., Reynolds, H. M. and Young, J. W. (1975) 'Investigation of inertial properties of the human body', AMRL T R 74–137, AD-A016–485, DOT-HS-801–430. Wright-Patterson Air Force Base, Ohio.

Chapman, A. E. (1985) 'The mechanical properties of human muscle', in R. L. Terjung (ed.), *Exercise and Sport Sciences Reviews – Volume 13*, pp. 443–502. London: MacMillan.

Cheng, K. B. and Hubbard, M. (2004) 'Optimal jumping strategies from compliant surfaces: a simple model of springboard standing jumps', *Human Movement Science*, 23: 35–48.

Clarys, J. P. and Marfell-Jones, M. J. (1986) 'Anthropometric prediction of component tissue masses in the minor limb segments of the human body', *Human Biology*, 58: 761–769.

Conceição, F., King, M. A., Yeadon, M. R. and Lewis, M. G. C. and Forrester, S. E. (2012) 'An isovelocity dynamometer method to determine mono-articular and bi-articular muscle parameters', *Journal of Applied Biomechanics*, 28: 751–760.

Dapena, J. (1981) 'Simulation of modified human airborne movements', *Journal of Biomechanics*, 14: 81–89.

Dempster, W. T. (1955) 'Space requirements of the seated operator', Wright Air Development Centre, Wright-Patterson Air Force Base, Ohio WADC-TR: 55–159.

De Zee, M. and Voigt, M. (2001) 'Moment dependency of the series elastic stiffness in the human plantar flexors measured in vivo', *Journal of Biomechanics*, 34: 1399–1406.

Dudley, G. A., Harris, R. T., Duvoisin, M. R., Hather, B. M. and Buchanan, P. (1990) 'Effect of voluntary vs artificial activation on the relationship of muscle torque to speed', *Journal of Applied Physiology*, 69: 2215–2221.

Edman, K. A. P. (1992) 'Contractile performance of skeletal muscle fibres', in P. V. Komi (ed.), *Strength and Power in Sport. Vol. III of the Encyclopedia of Sports Medicine*, pp. 114–133. Oxford: Blackwell Scientific.

Forrester, S. E., Yeadon, M. R., King, M. A. and Pain, M. T. G. (2011) 'Comparing different approaches for determining joint torque parameters from isovelocity dynamometer measurements', *Journal of Biomechanics*, 44: 955–961.

Fujii, N. and Hubbard, M. (2002) 'Validation of a three-dimensional baseball pitching model', *Journal of Applied Biomechanics*, 18: 135–154.

Gerritsen, K. G. M., van den Bogert, A. J. and Nigg, B. M. (1995) 'Direct dynamics simulation of the impact phase in heel – toe running', *Journal of Biomechanics*, 28: 661–668.

Gervais, P. (1994) 'A prediction of an optimal performance of the handspring 1½ front salto longhorse vault', *Journal of Biomechanics*, 27: 67–75.

Glynn, J. A., King, M. A. and Mitchell, S. R. (2011) 'A computer simulation model of tennis racket/ball impacts', *Sports Engineering*, 13: 65–72.

Goffe, W. L., Ferrier, G. D. and Rogers, J. (1994) 'Global optimisation of statistical functions with simulated annealing', *Journal of Econometrics*, 60: 65–99.

Greig, M. P. and Yeadon, M. R. (2000) 'The influence of touchdown parameters on the performance of a high jumper', *Journal of Applied Biomechanics*, 16: 367–378.

Gruber, K., Ruder, H., Denoth, J. and Schneider, K. (1998) 'A comparative study of impact dynamics: wobbling mass model versus rigid body models', *Journal of Biomechanics*, 31: 439–444.

Harry, J. D., Ward, A. W., Heglund, N. C., Morgan, D. L. and McMahon, T. A. (1990) 'Cross-bridge cycling theories cannot explain high-speed lengthening behaviour in frog muscle', *Biophysical Journal*, 57: 201–208.

Hatze, H. (1980) 'A mathematical model for the computational determination of parameter values of anthropomorphic segments', *Journal of Biomechanics*, 13: 833–843.

Hatze, H. (1981a) 'A comprehensive model for human motion simulation and its application to the take-off phase of the long jump', *Journal of Biomechanics*, 14: 135–142.

Hatze, H. (1981b) 'Estimation of myodynamic parameter values from observations on isometrically contracting muscle groups', *European Journal of Applied Physiology*, 46: 325–338.

Hatze, H. (1983) 'Computerized optimization of sports motions: an overview of possibilities, methods and recent developments', *Journal of Sports Sciences*, 1: 3–12.

Hiley, M. J., Apostolidis, A. and Yeadon, M. R. (2008) 'Optimisation of high bar circling technique for consistent performance of a triple piked somersault dismount', *Journal of Biomechanics*, 41: 1730–1735.

Hiley, M. J. and Yeadon, M. R. (2001) 'Swinging around the high bar', *Physics Education*, 36(1): 14–17.

Hiley, M. J. and Yeadon, M. R. (2003a) 'Optimum technique for generating angular momentum in accelerated backward giant circles prior to a dismount', *Journal of Applied Biomechanics*, 19: 119–130.

Hiley, M. J. and Yeadon, M. R. (2003b) 'The margin for error when releasing the high bar for dismounts', *Journal of Biomechanics*, 36: 313–319.

Hiley, M. J. and Yeadon, M. R. (2005) 'Maximal dismounts from high bar', *Journal of Biomechanics*, 38: 2221–2227.

Hiley, M. J. and Yeadon, M. R. (2007) 'Optimisation of backward giant circle technique on the asymmetric bars', *Journal of Applied Biomechanics*, 23: 301–309.

Hiley, M. J. and Yeadon, M. R. (2012a) 'The effect of cost function on optimum technique of the undersomersault on parallel bars', *Journal of Applied Biomechanics*, 28: 10–19.

Hiley, M. J. and Yeadon, M. R. (2012b) 'Achieving consistent performance in a complex whole body movement: the Tkatchev on high bar', *Human Movement Science*, 31: 834–843.

Hiley, M. J. and Yeadon, M. R. (2013) 'Investigating optimal technique in a noisy environment: application to the upstart on uneven bars', *Human Movement Science*, 32: 181–191.

Hiley, M. J., Zuevsky, V. and Yeadon, M. R. (2013) 'Is skilled technique characterised by high or low variability – an analysis of high bar giant circles?' *Human Movement Science*, 32: 171–180.

Hill, A. V. (1938) 'The heat of shortening and the dynamic constants of muscle', *Proceedings of the Royal Society Series B*, 126: 136–195.

Hinrichs, R. N. (1985) 'Regression equations to predict segmental moments of inertia from anthropometric measurements: an extension of the data of Chandler et al. (1975)', *Journal of Biomechanics*, 18: 621–624.

Hubbard, M. (1980) 'Human control of the skateboard', *Journal of Biomechanics*, 13: 745–754.

Hubbard, M. (1984) 'Optimal javelin trajectories', *Journal of Biomechanics*, 17: 777–787.

Hubbard, M. (1993) 'Computer simulation in sport and industry', *Journal of Biomechanics*, 26(1): 53–61.

Hubbard, M. and Alaways, L. W. (1987) 'Optimum release conditions for the new rules javelin', *International Journal of Sport Biomechanics*, 3: 207–221.

Hubbard, M. and Alaways, L. W. (1989) 'Rapid and accurate estimation of release conditions in the javelin throw', *Journal of Biomechanics*, 22: 583–596.

Hubbard, M., de Mestre, N. J. and Scott, J. (2001) 'Dependence of release variables in the shot put', *Journal of Biomechanics*, 34: 449–456.

Hubbard, M., Hibbard, R. L., Yeadon, M. R. and Komor, A. (1989) 'A multisegment dynamic model of ski jumping', *International Journal of Sport Biomechanics*, 5: 258–274.

Hubbard, M. and Trinkle, J. C. (1985a) 'Clearing maximum height with constrained kinetic energy', *ASME Journal of Applied Mechanics*, 52: 179–184.

Hubbard, M. and Trinkle, J. C. (1985b) 'Optimal Fosbury-flop high jumping', in D. A. Winter, R. W. Norman, R. P. Wells, K. C. Hayes and A. E. Patla (eds), *Biomechanics IX-B*, pp. 308–312. Champaign, IL: Human Kinetics.

Imura, A. and Yeadon, M. R. (2010) 'Mechanics of the Fouetté turn', *Human Movement Science*, 29: 947–955.

Jackson, M. I., Hiley, M. J. and Yeadon, M. R. (2011) 'A comparison of Coulomb and pseudo-Coulomb friction implementations: application to the table contact phase of gymnastics vaulting', *Journal of Biomechanics*, 44: 2706–2711.

Jensen, R. K. (1978) 'Estimation of the biomechanical properties of three body types using a photogrammetric method', *Journal of Biomechanics*, 11: 349–358.

Kentel, B. B., King, M. A., Mitchell, S. R. (2011) 'Evaluation of a subject-specific, torque-driven computer simulation model of one-handed tennis backhand groundstrokes', *Journal of Applied Biomechanics*, 27: 345–354.

King, A. I. (1984) 'A review of biomechanical models', *Journal of Biomechanical Engineering*, 106: 97–104.

King, M. A., Glynn, J. A. and Mitchell, S. R. (2011) 'Subject-specific computer simulation model for determining elbow loading in one-handed tennis backhand groundstrokes', *Sports Biomechanics*, 10: 391–406.

King, M. A., Kentel, B. B. and Mitchell, S. R. (2012b) 'The effects of ball impact location and grip tightness on the arm, racquet and ball for one-handed tennis backhand groundstrokes', *Journal of Biomechanics*, 45: 1048–1052.

King, M. A., Kong, P. W. and Yeadon, M. R. (2009) 'Determining effective subject-specific strength levels for forward dives using computer simulations of recorded performances', *Journal of Biomechanics*, 42: 2672–2677.

King, M. A., Lewis, M. G. C., Yeadon, M. R. (2012a) 'Is it necessary to include biarticular effects within joint torque representations of knee flexion and knee extension?' *International Journal for Multiscale Computational Engineering*, 10: 117–130.

King, M. A., Wilson, C. and Yeadon, M. R. (2003) 'Determination of knee joint moments during running jumps using a constrained forward dynamics simulation model', *International Journal of Computer Science in Sport*, 2: 102–103.

King, M. A., Wilson, C. and Yeadon, M. R. (2006) 'Evaluation of a torque-driven computer simulation model of jumping for height', *Journal of Applied Biomechanics*, 22: 264–274.

King, M. A. and Yeadon, M. R. (2002) 'Determining subject specific torque parameters for use in a torque driven simulation model of dynamic jumping', *Journal of Applied Biomechanics*, 18: 207–217.

King, M. A. and Yeadon, M. R. (2003) 'Coping with perturbations to a layout somersault in tumbling', *Journal of Biomechanics*, 36: 921–927.

King, M. A. and Yeadon, M. R. (2004) 'Maximising somersault rotation in tumbling', *Journal of Biomechanics*, 37: 471–477.

King, M. A. and Yeadon, M. R. (2005) 'Factors influencing performance in the Hecht vault and implications for modelling', *Journal of Biomechanics*, 38: 145–151.

King, M. A. and Yeadon, M. R. (2015) 'Advances in the development of whole body computer simulation modelling of sports technique', *Movement and Sport Sciences – Science and Motricité*, 90: 55–67.

King, M. A., Yeadon, M. R. and Kerwin, D. G. (1999) 'A two segment simulation model of long horse vaulting', *Journal of Sports Sciences*, 17: 313–324.

Koh, M., Jennings, L., Elliott, B. and Lloyd, D. (2003) 'A predicted optimal performance of the Yurchenko layout vault in women's artistic gymnastics', *Journal of Applied Biomechanics*, 19: 187–204.

Kong, P. W., Yeadon, M. R. and King, M. A. (2005) 'Optimisation of take-off techniques for maximum forward rotation in springboard diving', in Q. Wang (ed.), *XXIII*

International Symposium on Biomechanics in Sports, pp. 569–572. Beijing: The China Institute of Sport Science.

Lewis, M. G. C., King, M. A., Yeadon, M. R. and Conceição, F. (2012) 'Are joint torque models limited by an assumption of monoarticularity?' *Journal of Applied Biomechanics*, 28: 520–529.

MacKenzie, S. J. (2012) 'Club position relative to the golfer's swing plane meaningfully affects swing dynamics', *Sports Biomechanics*, 11: 149–164.

Martin, P. E., Mungiole, M., Marzke, M. W. and Longhill, J. M. (1989) 'The use of magnetic resonance imaging for measuring segment inertial properties', *Journal of Biomechanics*, 22: 367–369.

Miller, D. I. (1971) 'A computer simulation of the airborne phase of diving', in J. M. Cooper (ed.), *Selected Topics on Biomechanics*, pp. 207–215. Chicago, IL: Athletic Institute.

Miller, D. I. (1975) 'Computer simulation of human motion', in D. W. Grieve, D. I. Miller, D. Mitchelson, J. Paul and A. J. Smith (eds), *Techniques for the Analysis of Human Movement*, pp. 69–105. London: Lepus Books.

Mills, C., Pain, M. T. G. and Yeadon, M. R. (2006) 'Modelling a viscoelastic gymnastics landing mat during impact', *Journal of Applied Biomechanics*, 22: 103–111.

Mills, C., Pain, M. T. G. and Yeadon, M. R. (2008) 'The influence of simulation model complexity on the estimation of internal loading in gymnastics landings', *Journal of Biomechanics*, 41: 620–628.

Mills, C., Pain, M. T. G. and Yeadon, M. R. (2009) 'Reducing ground reaction forces in gymnastics' landings may increase internal loading', *Journal of Biomechanics*, 42: 671–678.

Mills, C., Yeadon, M. R. and Pain, M. T. G. (2010) 'Modifying landing mat material properties may decrease peak contact forces but increase forefoot forces in gymnastics landings', *Sports Biomechanics*, 9: 153–164.

Muramatsu, T., Muraoka, T., Takeshita, D., Kawakami, Y., Hirano, Y. and Fukunaga, T. (2001) 'Mechanical properties of tendon and aponeurosis of human gastrocnemius muscle in vivo', *Journal of Applied Physiology*, 90: 1671–1678.

Nagano, A. and Gerritsen, G. M. (2001) 'Effects of neuromuscular strength training on vertical jumping performance – a computer simulation study', *Journal of Applied Biomechanics*, 17: 113–128.

Nelder, J. A. and Mead, R. (1965) 'A simplex method for function minimisation', *Computer Journal*, 7: 308–313.

Neptune, R. R. and Hull, M. L. (1998) 'Evaluation of performance criteria for simulation of submaximal steady-state cycling using a forward dynamic model', *Journal of Biomechanical Engineering*, 120: 334–341.

Neptune, R. R. and Hull, M. L. (1999) 'A theoretical analysis of preferred pedaling rate selection in endurance cycling', *Journal of Biomechanics*, 32: 409–415.

Neptune, R. R. and Kautz, S. A. (2000) 'Knee joint loading in forward versus backward pedaling: implications for rehabilitation strategies', *Clinical Biomechanics*, 15: 528–535.

Niklas, K. J. (1992) *Plant Biomechanics: An Engineering Approach to Plant Form and Function*, Chicago, IL: University of Chicago Press.

Pain, M. T. G. and Challis, J. H. (2001a) 'High resolution determination of body segment inertial parameters and their variation due to soft tissue motion', *Journal of Applied Biomechanics*, 17: 326–334.

Pain, M. T. G. and Challis, J. H. (2001b) 'The role of the heel pad and shank soft tissue during impacts: a further resolution of a paradox', *Journal of Biomechanics*, 34: 327–333.

Pain, M. T. G. and Challis, J. H. (2006) 'The influence of soft tissue movement on ground reaction forces, joint torques and joint reaction forces in drop landings', *Journal of Biomechanics*, 39: 119–124.

Panjabi, M. (1979) 'Validation of mathematical models', *Journal of Biomechanics*, 12: 238.

Press, W. H., Flannery, B. P., Teukolsky, S. A. and Vetterling, W. T. (1988) *Numerical Recipes. The Art of Scientific Computing*, Cambridge: Cambridge University Press.

Requejo, P. S., McNitt-Gray, J. L. and Flashner, H. (2004) 'Modification of landing conditions at contact via flight phase control', *Biological Cybernetics*, 90: 327–336.

Sawicki, G. S., Hubbard, M. and Stronge, W. J. (2003) 'How to hit home runs: optimum baseball bat swing parameters for maximum range trajectories', *American Journal of Physics*, 71: 1152–1162.

Schwark, B. N., Mackenzie, S. J. and Sprigings, E. J. (2004) 'Optimizing the release conditions for a free throw in wheelchair basketball', *Journal of Applied Biomechanics*, 20: 153–166.

Sheets, A. L. and Hubbard, M. (2008) 'Evaluation of a subject-specific female gymnast model and simulation of an uneven parallel bar swing', *Journal of Biomechanics*, 41, 3139–3144.

Sprigings, E. J., Lanovaz, J. L., Watson, L. G. and Russell, K. W. (1998) 'Removing swing from a handstand on rings using a properly timed backward giant circle: a simulation solution', *Journal of Biomechanics*, 31: 27–35.

Sprigings, E. J. and Miller, D. I. (2004) 'Optimal knee extension timing in springboard and platform dives from the reverse group', *Journal of Applied Biomechanics*, 20: 244–252.

Sprigings, E. J. and Yeadon, M. R. (1997) 'An insight into the reversal of rotation in the Hecht vault', *Human Movement Science*, 16: 517–532.

Tan, J. C. C. (1997) 'The mechanics of the curved approach in high jumping', *Unpublished Doctoral Dissertation*, Loughborough University, UK.

van den Bogert, A. J. (1994) 'Optimization of the human engine: application to sprint cycling', in W. Herzog, B. Nigg and T. van den Bogert (eds), *Canadian Society for Biomechanics: Proceedings of the Eight Biennial Conference and Symposium*, pp. 160–162. Calgary: Canadian Society for Biomechanics.

van den Bogert, A. J. and Ackermann, M. (2009) 'Direct collocation for simulation and optimization of movement', in Proceedings of the 12th International Symposium on Computer Simulation in Biomechanics, Cape Town: International Society of Biomechanics.

van der Helm, F. C. T. (1994) 'A finite element musculoskeletal model of the shoulder mechanism', *Journal of Biomechanics*, 27: 551–569.

van Gheluwe, B. (1981) 'A biomechanical simulation model for airborne twist in backward somersaults', *Journal of Human Movement Studies*, 7: 1–22.

van Soest, A. J., Bobbert, M. F. and Ingen Schenau, G. J. van (1994) 'A control strategy for the execution of explosive movements from varying starting positions', *Journal of Neurophysiology*, 71: 1390–1402.

van Soest, A. J. and Casius, L. J. R. (2003) 'The merits of a parallel genetic algorithm in solving hard optimization problems', *Journal of Biomechanical Engineering*, 125: 141–146.

Vaughan, C. L. (1984) 'Computer simulation of human motion in sports biomechanics', in R. L. Terjung (ed.), *Exercise and Sport Sciences Reviews – Volume 12*, pp. 373–398. New York: Palgrave Macmillan.

Vaughan, C. L. (ed.) (1989) *Biomechanics of Sport*, Boca Raton, FL: CRC Press.

Westing, S. H., Seger, J. Y., Karlson, E. and Ekblom, B. (1988) 'Eccentric and concentric torque-velocity characteristics of the quadriceps femoris in man', *European Journal of Applied Physiology*, 58: 100–104.

Wilson, C., King, M. A. and Yeadon, M. R. (2006) 'Determination of subject-specific model parameter visco-elastic elements', *Journal of Biomechanics*, 39: 1883–1890.

Wilson, C., King, M. A. and Yeadon, M. R. (2011) 'The effects of initial conditions and takeoff technique on running jumps for height and distance', *Journal of Biomechanics*, 44: 2207–2212.

Wilson, C., Yeadon, M. R. and King, M. A. (2007) 'Considerations that affect optimised simulation in a running jump for height', *Journal of Biomechanics*, 40: 3155–3161.

Wright, I. C., Neptune, R. R., van den Bogert, A. J. and Nigg, B. M. (1998) 'Passive regulation of impact forces in heel – toe running', *Clinical Biomechanics*, 13: 521–531.

Yeadon, M. R. (1987) 'Theoretical models and their application to aerial movement', in J. Atha and B. van Gheluwe (eds), *Medicine and Sport Science, 25: Current Research in Sports Biomechanics*, pp. 86–106. Basel: Karger.

Yeadon, M. R. (1989) 'Twisting techniques used in freestyle aerial skiing', *International Journal of Sport Biomechanics*, 5: 275–284.

Yeadon, M. R. (1990a) 'The simulation of aerial movement – part I: the determination of orientation angles from film data', *Journal of Biomechanics*, 23: 59–66.

Yeadon, M. R. (1990b) 'The simulation of aerial movement – part II: a mathematical inertia model of the human body', *Journal of Biomechanics*, 23: 67–74.

Yeadon, M. R. (1993a) 'The biomechanics of twisting somersaults. Part I: Rigid body motions', *Journal of Sports Sciences*, 11: 187–198.

Yeadon, M. R. (1993b) 'The biomechanics of twisting somersaults. Part II: Contact twist', *Journal of Sports Sciences*, 11: 199–208.

Yeadon, M. R. (1993c) 'The biomechanics of twisting somersaults. Part III: Aerial twist', *Journal of Sports Sciences*, 11: 209–218.

Yeadon, M. R. (1993d) 'The biomechanics of twisting somersaults. Part IV: Partitioning performance using the tilt angle', *Journal of Sports Sciences*, 11: 219–225.

Yeadon, M. R. (1993e) 'Twisting techniques used by competitive divers', *Journal of Sports Sciences*, 11: 337–342.

Yeadon, M. R. (1994) 'Twisting techniques used in dismounts from rings', *Journal of Applied Biomechanics*, 10: 178–188.

Yeadon, M. R. (1997) 'Twisting double somersault high bar dismounts', *Journal of Applied Biomechanics*, 13: 76–87.

Yeadon, M. R. (2013) 'The limits of aerial twisting techniques in the aerials event of freestyle skiing', *Journal of Biomechanics*, 46: 1008–1013.

Yeadon, M. R., Atha, J. and Hales, F. D. (1990a) 'The simulation of aerial movement – IV: a computer simulation model', *Journal of Biomechanics*, 23: 85–89.

Yeadon, M. R. and Brewin, M. A. (2003) 'Optimised performance of the backward longswing on rings', *Journal of Biomechanics*, 36: 545–552.

Yeadon, M. R. and Challis, J. H. (1994) 'The future of performance related sports biomechanics research', *Journal of Sports Sciences*, 12: 3–32.

Yeadon, M. R. and Hiley, M. J. (2000) 'The mechanics of the backward giant circle on the high bar', *Human Movement Science*, 19: 153–173.

Yeadon, M. R. and Hiley, M. J. (2014) 'The control of twisting somersaults', *Journal of Biomechanics*, 47: 1340–1347.

Yeadon, M. R. and King, M. A. (2002) 'Evaluation of a torque driven simulation model of tumbling', *Journal of Applied Biomechanics*, 18: 195–206.

Yeadon, M. R., King, M. A., Forrester, S., Caldwell, G. E. and Pain, M. T. G. (2010) 'The need for muscle co-contraction prior to a landing', *Journal of Biomechanics*, 43: 364–369.

Yeadon, M. R., King, M. A. and Wilson, C. (2006b) 'Modelling the maximum voluntary joint torque/angular velocity relationship in human movement', *Journal of Biomechanics*, 39: 476–482.

Yeadon, M. R. and Knight, J. P. (2012) 'A virtual environment for learning to view during aerial movements', *Computer Methods in Biomechanics and Biomedical Engineering*, 15: 919–924.

Yeadon, M. R., Kong, P. W. and King, M. A. (2006a) 'Parameter determination for a computer simulation model of a diver and a springboard', *Journal of Applied Biomechanics*, 22: 167–176.

Yeadon, M. R., Lee, S. and Kerwin, D. G. (1990b) 'Twisting techniques used in high bar dismounts', *International Journal of Sport Biomechanics*, 6: 139–146.

Yeadon, M. R. and Mikulcik, E. C. (1996) 'The control of non-twisting somersaults using configurational changes', *Journal of Biomechanics*, 29: 1341–1348.

Yeadon, M. R. and Morlock, M. (1989) 'The appropriate use of regression equations for the estimation of segmental inertia parameters', *Journal of Biomechanics*, 22: 683–689.

Yeadon, M. R. and Trewartha, G. (2003) 'Control strategy for a hand balance', *Motor Control*, 7: 411–430.

Zatsiorsky, V. M. (2002) *Kinetics of Human Motion*, Champaign, IL: Human Kinetics.

Zatsiorsky, V. M., Seluyanov, V. N. and Chugunova, L. (1990a) 'In vivo body segment inertial parameters determination using a gamma-scanner method', in N. Berme and A. Cappozzo (eds), *Biomechanics of Human Movement: Applications in Rehabilitation, Sports and Ergonomics*, pp. 186–202. Worthington, OH: Bertec Corporation.

Zatsiorsky, V. M., Seluyanov, V. N. and Chugunova, L. G. (1990b) 'Methods of determining mass-inertial characteristics of human body segments', in G. G. Chernyi and S. A. Regirer (eds), *Contemporary Problems of Biomechanics*, pp. 272–291. USA: CRC Press.

INDEX

Page numbers in italic indicate a figure on the corresponding page.